Extremals for the Sobolev Inequality and the Quaternionic Contact Yamabe Problem

Extremals for the
Sobolev Inequality and
the Quaternionic Contact
Yamabe Problem

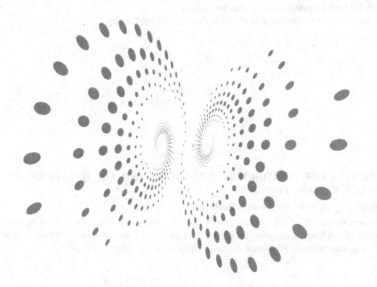

Stefan P Ivanov
University of Sofia "St Kliment Ohridski", Bulgaria

Dimiter N Vassilev
The University of New Mexico, USA

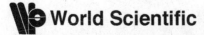 **World Scientific**

NEW JERSEY · LONDON · SINGAPORE · BEIJING · SHANGHAI · HONG KONG · TAIPEI · CHENNAI

Published by

World Scientific Publishing Co. Pte. Ltd.

5 Toh Tuck Link, Singapore 596224

USA office: 27 Warren Street, Suite 401-402, Hackensack, NJ 07601

UK office: 57 Shelton Street, Covent Garden, London WC2H 9HE

British Library Cataloguing-in-Publication Data
A catalogue record for this book is available from the British Library.

ISBN-13 978-981-4295-70-3
ISBN-10 981-4295-70-1

Printed in Singapore.

Preface

These lecture notes grew out of our work on the quaternionic contact Yamabe problem and in particular the cases of the quaternionic Heisenberg group and the quaternionic contact sphere. Similarly to its Riemannian and CR counterparts the complete understanding of the problem requires a number of beautiful and powerful techniques. Unlike the Riemannian and CR cases, the solution of the quaternionic contact Yamabe problem has yet to be achieved. Nevertheless, our efforts have been rewarded with a complete solution on the seven dimensional quaternionic contact sphere and quaternionic Heisenberg group. We have also determined the quaternionic contact Yamabe constant of the quaternionic contact sphere of any dimension and all functions achieving this constant. More importantly, the quaternionic contact Yamabe problem led to some new ideas and geometric structures that are interesting on their own. Some of these developments will find a place in this book, such as the quaternionic contact conformal tensor and the qc-Einstein structures since both are crucial for the solution of the quaternionic contact Yamabe problem. The book presupposes the paper [94] from which we borrow a number of fundamental results without supplying their proofs. On the other hand, we leave out completely some results of [94] concerning anti-regular, quaternionic pluriharmonic and anti-CRF functions (see also [154], [156], [138; 139], [140], [43], [44], [123], [12], [86], [5; 6; 7], [8]), pseudo-Einstein strictures, and infinitesimal automorphisms of quaternionic contact manifolds.

The Yamabe problem arouse from two questions, one coming from the area of geometry, the other from the area of analysis. Yamabe [174] considered conformal transformations, i.e., transformations that preserve angles, of a Riemannian metric on a given compact manifold M to one with a constant scalar curvature. The existence of such conformal deformation is equivalent to the existence of a positive solution of a non-linear partial differential equation - the Yamabe equation,

$$4\frac{n-1}{n-2}\triangle u + Su = -\tilde{S}u^{2^*-1}, \qquad u \in \mathcal{C}^\infty, \qquad u \geq 0, \qquad (0.1)$$

where n is the dimension of M, $\triangle$ is the Laplace operator of the fixed Riemannian metric g, S and $\tilde{S} = const$ are the scalar curvatures of g and $\tilde{g} = u^{2^*-2} g$ respectively, and $2^* = 2n/(n-2)$. Any positive solution to (0.1) gives a Yamabe metric, i.e., a metric of constant scalar curvature in the fixed conformal class $[g]$. Yamabe [174] observed that (0.1) is (up to a scaling) the Euler-Lagrange equation of the (Yamabe) functional defined by

$$\Upsilon(u) = \frac{\int_M \left(4\frac{n-1}{n-2} |\nabla u|^2 + \mathrm{Scal}\, u^2\right) dv_g}{\left(\int_M u^{2^*} dv_g\right)^{2/2^*}}$$

which led him to consider the infimum of this functional thereby defining the so called Yamabe constant of the conformal class

$$\lambda(M) \stackrel{def}{=} \lambda(M, [g]) = \inf\{\Upsilon(u) \, : \, u \in \mathcal{C}^\infty(M), \, u > 0\}.$$

Yamabe's goal was to prove the existence of metrics of constant scalar curvature conformal to a given metric g by showing that the Yamabe constant is achieved, see also [163]. We shall refer to the question of the precise value of this constant and the metrics in the fixed conformal class for which it is achieved as the *Yamabe constant problem*. As well known, in general, there are Yamabe metrics of higher energy that do not realize the Yamabe constant.

When the ambient space is the Euclidean space $\mathbb{R}^n$, G. Talenti [159] and T. Aubin [14], [16] described *all positive solutions* of a more general equation

$$\triangle_p u = \sum_{j=1}^n \partial_j(|\nabla u|^{p-2}\partial_j u) = -|u|^{p^*-2}u, \qquad u \in \mathcal{C}^\infty(\mathbb{R}^n), \qquad (0.2)$$

that is, the Euler-Lagrange equation associated to the functional defining the best constant, i.e., the norm, of the L^p Sobolev embedding theorem

$$\left(\int_{\mathbb{R}^n} |u|^{p^*} dx\right)^{1/p^*} \leq S_p \left(\int_{\mathbb{R}^n} |\nabla u|^p dx\right)^{1/p}, \qquad (0.3)$$

holding for any $u \in C_o^\infty(\mathbb{R}^n)$, where $p^* = np/(n-p)$ is the so-called Sobolev conjugate to $1 < p < n$. Remarkably, all of these solutions turned the Sobolev inequality in an equality, hence we can determine the value of the smallest constant S_p for which the equality in (0.3) holds true. Of course, this was Talenti's motivation to consider (0.2).

In the case of the standard sphere, using the stereographic projection - an example of a conformal transformation, the description of all Yamabe metrics on the round sphere is equivalent to the above question on $\mathbb{R}^n$ in the L^2 case ($p = 2$). The solution of this special case is an important step in solving the general Yamabe problem, the solution of which in the case of a compact Riemannian manifold was completed in the 80's after the work of T. Aubin [14; 15; 16; 17] and R. Schoen [149], the latter using the positive mass theorem of R. Schoen and S. T. Yau, see [122] for references and history, see also [19].

More recently the CR Yamabe problem was studied, which is the problem of obtaining and classifying all contact forms of constant Webster-Tanaka scalar curvature within a fixed conformal class of contact forms compatible with the given CR structure. The solution of the existence part of the general CR problem is now complete in the case of a compact CR manifold after the works of D. Jerison and J. Lee, [101] - [104], and N. Gamara and R. Yacoub, [72], [73]. Similarly to the Riemannian case, all contact forms conformal to the standard CR pseudo-hermitian structure on the $(2n+1)$-dimensional sphere in $\mathbb{C}^n$ that are of constant Tanaka-Webster scalar curvature turn out to be minimizers of the corresponding CR Yamabe functional. The latter is defined with the help of the fixed contact form θ with Tanaka-Webster scalar curvature S as follows, [101],

$$\Upsilon(u) = \frac{\int_M \left(\frac{2n+2}{n} |\nabla u|_\theta^2 + S u^2\right) \theta \wedge (d\theta)^n}{\left(\int_M u^{2^*} \theta \wedge (d\theta)^n\right)^{2/2^*}}, \qquad 2^* = 2n+2,$$

with Yamabe constant given by

$$\lambda(M) \stackrel{def}{=} \lambda(M, [\theta]) = \inf\{\Upsilon(u) : u \in \mathcal{C}^\infty(M), \ u > 0\}.$$

The case of the odd dimensional spheres equipped with the standard CR structures turns out to be equivalent to the problem of determining the best constant in the L^2 Folland-Stein Sobolev type inequality on the Heisenberg group and the functions for which equality is achieved.

Regarding the Folland-Stein inequality, it is known that on any Carnot group G there is the following inequality, [65] and [67]. For any $1 < p < Q$, where Q is the homogeneous dimension of the manifold G, define the Sobolev conjugate exponent $p^* = \frac{pQ}{Q-p}$. There exists a constant $S_p = S_p(G) > 0$ such that for any $u \in C_o^\infty(G)$

$$\left(\int_G |u(g)|^{p^*} \, dH\right)^{1/p^*} \leq S_p \left(\int_G |Xu(g)|^p \, dH\right)^{1/p}, \tag{0.4}$$

where Xu denotes the horizontal gradient involving only the derivatives along the first layer of the Lie algebra of G and dH is a Haar measure on the group.

Very important examples of Carnot groups are given by the so-called Heisenberg type groups. Among them there are the Iwasawa type groups, which arise as the nilpotent part N in the Iwasawa decomposition $G = KAN$ associated to a simple Lie group G of real rank one. From a different point of view the Iwasawa groups are the ideal boundaries of the symmetric spaces of non-compact type of real rank one, see [49] and also [24] where the geometric structure on the boundary arises as a conformal infinity of the hyperbolic metric. In the subsequent chapters we shall focus on the case of the quaternionic Heisenberg group, which together with the standard and the octonionic Heisenberg gives all groups of Iwasawa type. We shall be more explicit at the appropriate place, but for the moment it suffices to say that the quaternionic case is the main subject of these lecture notes.

The Folland-Stein inequality is a generalization of the Sobolev inequality for functions on $\mathbb{R}^n$ and thus plays a fundamental role in the analysis on Carnot groups. Denote by S_p the smallest possible constant for which the equality in (0.4) holds. This is the so-called best constant in the Folland-Stein inequality and the functions achieving the equality in (0.4) are called sometimes minimizers. The precise value of the best constant is unknown. A more modest problem is to determine S_2 and the minimizers in the case of groups of Iwasawa type. The above mentioned results of Talenti, Aubin, and Jerison and Lee solve this problem, correspondingly, in the degenerate Iwasawa group case of Euclidean space and the Heisenberg group. In either case, it is shown that the extremals of the L^2 Sobolev type inequality give, in fact, all non-negative solutions of the Yamabe equation on the group. On the quaternionic Heisenberg group of dimension seven this was achieved in [95]. Furthermore, in [96] following the method of [70], the best constant in the L^2 Folland-Stein inequality and the functions for which it is achieved were determined on the quaternionic Heisenberg group of any dimension. On any Carnot group, the Yamabe equation is the non-linear second order sub-elliptic equation

$$\mathcal{L}u = -u^{\frac{Q+2}{Q-2}}, \qquad u \in \overset{o}{\mathcal{D}}{}^{1,2}(G), \qquad u \geq 0, \qquad (0.5)$$

where $\mathcal{L}$ is the sub-Laplacian on the group and Q is the homogeneous dimension of G. It is, up to a scaling, the Euler-Lagrange equation of the variational problem associated with S_2 (i.e., the L^2 case), hence its positive solutions are critical points of the natural functional. One consequence of

[95] is that the solutions of (0.5) on the seven dimensional quaternionic Heisenberg group have a certain symmetry, so-called cylindrical symmetry. The classification of the cylindrically symmetric solutions of the Yamabe equation on groups of Iwasawa type, not just on the quaternionic Heisenberg group, has been done in [76]. In the case of $\mathbb{R}^n$, the Yamabe equation (0.5) admits up to a translation only radial solutions, see [159] and also [78]. In the case of the Heisenberg group the solutions are known to be biradial after a suitable translation [103]. Proving symmetry (geometric) properties of solutions of sub-elliptic equations is currently out of reach. The method of the so-called moving plane was used in [76] but it worked only in directions along the center of the group. Currently the only available method of determining the positive entire solutions of the Yamabe equation in the setting of groups of Iwasawa type is to invoke the relevant geometry which helps in finding a very non-trivial analogue of the divergence formula first used by M. Obata [133]. The latter exploited the Riemannian geometry and the Levi-Civita connection of the Riemannian metric. This was also the approach taken by D. Jerison and J. Lee [103] who found a divergence formula in terms of the Tanaka-Webster connection. In either case, the divergence formula is used to see that the "new" Yamabe metric is also Einstein, the latter notion interpreted appropriately. On the other hand, functions for which the Sobolev inequality becomes an equality can be found with symmetrization arguments. In the Euclidean case Talenti [159] used the spherical symmetric rearrangement to reduce the problem to an ordinary differential equation, see also Aubin [14] - [16]. Very recently, another symmetrization argument was proposed in [30] and [70] based on Szegö and Hersch's center of mass method [158] and [89]. This method yields not only the best constant but also all minimizers in the Euclidean and Heisenberg group versions of the L^2 Sobolev type inequality. We shall apply this method successfully also on the quaternionic Heisenberg group in Theorem 6.1.1 based on [96]. Again, we exploit the conformal nature of the problem and reduce it to the corresponding Yamabe constant problem on the quaternionic contact sphere where the actual symmetrization argument is used.

In the second part of the book we consider the Yamabe problem on the quaternionic Heisenberg group and the quaternionic contact sphere. This problem turns out to be equivalent to the quaternionic contact Yamabe problem on the unit $(4n + 3)$-dimensional sphere in the quaternion space due to the quaternionic Cayley transform, which is a conformal quaternionic contact transformation. In the quaternionic setting, the relevant geometry was developed by O. Biquard [24]. Studying conformal boundaries at infinity of quaternionic hyperbolic spaces O. Biquard introduced in [24] the notion of quaternionic contact structure. In particular, a connection, referred to as the Biquard connection, suitable for handling geometric questions in geometries modeled on the quaternionic Heisenberg group was defined in [24]. The quaternionic contact Yamabe problem consists of finding in the conformal class of a given quaternionic contact structure one with a constant scalar curvature of the Biquard connection and possibly describing all such structures. Yet again, the question is of variational nature. Given a compact quaternionic contact manifold M with a fixed conformal class defined by a quaternionic contact form $[\eta]$ and volume form Vol_η (6.9) the Yamabe equation characterizes the non-negative extremals of the constrained Yamabe functional defined by (see also (6.2))

$$\Upsilon(u) = \int_M \left(4\frac{n+2}{n+1}\, |\nabla u|^2 + \text{Scal}\, u^2 \right) Vol_\eta, \quad \int_M u^{2^*} Vol_\eta = 1,\, u > 0.$$

According to [168], and similarly to the Riemannian and CR Yamabe problems, the quaternionic contact Yamabe constant

$$\lambda(M) \overset{def}{=} \lambda(M, [\eta]) = \inf\{\Upsilon(u) : \int_M u^{2^*} Vol_\eta = 1,\, u > 0\}$$

of a compact quaternionic contact manifold is always less or equal than that of the standard quaternionic contact sphere. Furthermore, if the constant is strictly less than that of the sphere the quaternionic contact Yamabe problem has a solution, i.e., there is a global quaternionic contact conformal transformation sending the given quaternionic contact structure to a quaternionic contact structure with constant quaternionic contact scalar curvature. For the seven dimensional quaternionic contact sphere or Heisenberg group we are able to describe all Yamabe quaternionic contact sctructures in the conformal class of the standard quaternionic contact structure on the seven dimensional sphere, see Theorem 6.1.2.

A natural conjecture is that the quaternionic contact Yamabe constant of every compact locally non-flat manifold (in conformal quaternionic contact sense) is strictly less than the quaternionic contact Yamabe constant of the sphere with its standard quaternionic contact structure. Guided by the conformal and CR cases, we expect that the conformal quaternionic contact curvature tensor (5.48) will be a useful tool in the analysis of the quaternionic contact Yamabe problem.

Before giving the brief description of each chapter we should mention that we consider $1 < p < Q$ rather than $1 \leq p < Q$, i.e., $p = 1$ is excluded. As well known, the case $p = 1$ (Gagliardo-Nireneberg inequality) is essentially the isoperimetric inequality and implies the validity of the Sobolev (type) inequality for $1 < p < Q$. However, when one is interested in the value of the best constant the cases $p = 1$ and $1 < p < Q$ are completely different. In addition, isoperimetric inequalities and profiles on Carnot groups will certainly take us beyond the scope of this book, but the reader can consult [164], [148], [85] and [36] for more details and references.

Chapter 1 contains the necessary results proving the existence Theorem 1.4.2 and $\mathcal{C}^\infty$ smoothness Theorem 1.6.9 of the functions achieving the best constant in (0.4). We devote a section for the case when only functions with partial symmetry are allowed. This is due to the limitations of the results of Chapter 3. The main results in this section come from [76], [165] and [166].

Chapter 2 is devoted to the groups of Heisenberg and Iwasawa types. The latter serve as the flat models for the Carnot-Caratheodory geometries central for the whole book. We present the definitions and properties of the Cayley and Kelvin transforms. We end the section with an explicit solution of the Yamabe equation on a group of Heisenberg type [75].

In Chapter 3 Theorem 3.5.1, following [75], we determine the solutions of the Yamabe equation on groups of Iwasawa type assuming a-priori that the solutions are partially cylindrical. In addition, using [167] in Theorem 3.6.2 we determine the extremals of some Euclidean Hardy-Sobolev inequalities involving the distance to a $(n - k)$-dimensional coordinate subspace of $\mathbb{R}^n$. We achieve the results as a direct consequence of [75] by relating extremals on the Heisenberg groups to extremals in the Euclidean setting. When $k = n$ the above inequality becomes the Caffarelli-Kohn-Nirenberg inequality, see [34], for which the optimal constant was found in [77].

In Chapter 4 we specialize to the quaternionic Heisenberg group and the geometry modeled on it - the quaternionic contact geometry. We start with a definition of a quaternionic contact structure and derive the Biquard connection in a way slightly different from [24]. Our motivation was to give a direct approach based on the existence of Reeb vector fields rendering the construction more explicit and accessible for the general audience. In addition, this will facilitate the numerous explicit calculation that are needed in the subsequent sections. An immediate consequence is the characterization Theorem 4.3.15 of quaternionic contact structures locally isomorphic to the quaternionic Heisenberg group. We proceed with the study of qc-Einstein structures and the quaternionc contact conformal transformations of quaternionic contact structures. In Proposition 4.3.8 we derive the structure equations of a quaternionic contact manifold and in Theorems 4.4.2

and 4.4.4 give a characterization of qc-Einstein structures using the fundamental four form defining the quaternionic contact structure.

In Chapter 5 we turn to the construction of a quaternionic contact conformally invariant curvature tensor, Definition 5.3.2, characterizing quaternionic contact structures which are locally quaternionic contact conformally equivalent to the standard flat quaternionic contact structure on the quaternionic Heisenberg group. We begin with a complete and slightly different than [97] proof of the existence and invariance of the quaternionic contact conformal curvature tensor discovered by the authors in [97]. We sketch the proof of the quaternionic contact flatness Theorem 5.3.5 leaving out some of the computational details, which can be found in [97].

Chapter 6 begins with a brief background from [94] and in particular Theorem 6.2.5 which determines all qc-Einstein structures conformal to the standard quaternionic contact structure on the quaternionic Heisenberg group. Then we turn to the divergence formula of [95] supplying the final step in the proof of Theorem 6.1.2 which gives explicitly *all solutions* of the quaternionic contact Yamabe equation on the *seven dimensional* quaternionic Heisenberg group and seven dimensional quaternionic contact sphere. We end the chapter with Theorem 6.1.1, based on [96], where we determine the sharp constant and all functions for which equality holds in the L^2 Folland-Stein inequality on the quaternionic Heisenberg group of *any* dimension. Equivalently, we determine all functions achieving the Yamabe constant of the conformal class of the standard quaternionic contact spheres. Nevertheless, this leaves open the quaternionic contact Yamabe problem on the standard quaternionic contact spheres of dimension greater than seven. Naturally, we expect that there are no other solutions besides those with lowest energy, i.e., the known functions achieving the Yamabe constant.

In Chapter 7 Theorem 7.3.5 we give a proof of the Cartan-Chern-Moser result following the approach we took to derive the quaternionic contact conformal curvature tensor thus presenting the new proof of the Cartan-Chern-Moser theorem outlined in [99]. The reader should be aware that the quaternionic contact case requires a considerable amount of calculations while the details in the CR case are far less and reading of the CR case before the quaternionic contact case might be useful.

Acknowledgments

It is a pleasure to acknowledge the many discussions and joint work with Nicola Garofalo, Ivan Minchev and Simeon Zamkovoy who influenced and contributed directly and indirectly in the writing of this book. Thanks are also due to many friends, especially Marisa Fernández, for discussions that helped in achieving a clearer presentation.

The authors would like to thank the various institutions, Sofia University, University of New Mexico, UC Riverside, Universidad del País Vasco, the Max-Planck-Institut für Mathematik - Bonn, and ICTP Trieste for the support and stimulating environment, the Bulgarian Science Fund for the partial support under Contract Idei - DO 02-257/18.12.2008 and DID 02-39/21.12.2009, and the Centre de Recherches Mathématiques and CIRGET, Montréal where the geometric part of the project initiated a few years ago.

Contents

PART I
Analysis

Chapter 1

Variational problems related to Sobolev inequalities on Carnot groups

1.1 Introduction

This chapter contains the core of the analysis side of the subject of the book. The main topic is the sharp form of the Sobolev type embedding theorem (1.1) established by Folland and Stein on Carnot groups, [65]. Let G be a Carnot group of homogeneous dimension Q and p a constant, $1 < p < Q$. There exists a constant $S_p = S_p(G) > 0$ such that for $u \in C_o^\infty(G)$

$$\left(\int_G |u|^{p^*} dH \right)^{1/p^*} \leq S_p \left(\int_G |Xu|^p dH \right)^{1/p}. \tag{1.1}$$

Here, the "horizontal" gradient, $|Xu|$, is defined by $|Xu| = (\sum_{j=1}^m (X_j u)^2)^{1/2}$, where $X = \{X_1, \ldots, X_m\}$ is a left invariant basis of the first layer V_1 whose commutators generate the whole algebra, $p^* = \frac{pQ}{Q-p}$ and dH is a fixed Haar measure on G. Unlike the Euclidean case, [159] and [14], the value of the best possible constant or the non-negative functions for which it is achieved is presently unknown.

We start the chapter with reviewing some necessary background on Carnot groups and their sub-Rimannian geometry in Section 1.2. The following Section 1.3 is devoted to the relevant Sobolev spaces, their duals, and weak topologies. After these two preliminary sections we turn to the question of existence of (extremals) minimizers in (1.1), i.e., functions for which inequality (1.1) turns into an equality. The difficulty in finding such functions stems from the fact that the considered embedding of Sobolev spaces is not compact and, crucially, there is a (noncompact) group of dilations preserving the set of extremals, see equation (1.29) and the paragraph following it. In particular, starting from an extremal we can construct a sequence of extremals which converge to the zero function. Thus, an argument proving the existence of an extremal by taking a sequence of func-

3

tions converging to an extremal will fail unless a more delicate analysis and modification (by scaling and translating) of the sequence is performed. The concentration-compactness principle of P. L. Lions, see [125], [126], will be applied in Section 1.4 in the homogeneous setting of a Carnot group G to prove that for any $1 < p < Q$ the best constant in the Folland-Stein embedding (1.1) is achieved. This method does not allow an explicit determination of the best constant or the functions for which it is achieved, but it allows us to see that for any $1 < p < Q$ the quasi-linear equation with critical exponent,

$$\mathcal{L}_p u \ = \ \sum_{j=1}^n X_j(|Xu|^{p-2}X_j u) \ = \ - u^{p^*-1} \qquad \text{in } \ G, \qquad (1.2)$$

possesses a weak non-negative solution, which is also, up to a constant, an extremal for the following variational problem,

$$I \ \stackrel{def}{=} \ \inf \left\{ \int_G |Xu|^p \ \mid \ u \in C_o^\infty(G), \ \int_G |u|^{p^*} = 1 \right\}. \qquad (1.3)$$

Here, we used that $\mathcal{L}_p(cu) = c^{p-1}u$ to reduce the equation given by the Euler-Lagrange multiplier to (1.2). In general, the norm of the embedding is achieved on the space $\overset{o}{\mathcal{D}}{}^{1,p}(G)$, which is the closure of $C_o^\infty(G)$ with respect to the norm

$$\|u\| \ \equiv \ \|u\|_{\overset{o}{\mathcal{D}}{}^{1,p}(G)} \ = \ \left(\int_G |Xu|^p dH \right)^{1/p}. \qquad (1.4)$$

We shall also consider a similar problem restricting the test functions to these with a certain symmetry, see Section 1.5. We prove that the corresponding infimum is achieved again.

In the last section of Chapter 1 we study the optimal regularity of solutions of the relevant non-linear sub-elliptic equations. Using Moser iteration arguments in Section 1.6.1 we show that any weak solution of the equation

$$\mathcal{L}_p u \ = \ \sum_{j=1}^n X_j(|Xu|^{p-2}X_j u) \ = \ - |u|^{p^*-2}u \qquad (1.5)$$

is a bounded function. We actually will consider a more general equation revealing the key properties that are needed in order for the proof to go through. We proceed to show that in the case $p = 2$ the weak solutions are smooth functions by employing sub-elliptic regularity estimates combined with the Hopf Lemma or the Harnack inequality.

As already mentioned, it is an open question to find the exact norm of the Folland-Stein embedding. A more tractable problem is obtained by requiring that the group be of Heisenberg (in fact Iwasawa) type. In the particular case of the Heisenberg group this problem was solved in the L^2 case, $p = 2$, in [103]. Later, in Chapter 3, we shall find the precise value of the norm of the embedding on any group of Iwasawa type when considering only functions with symmetries and $p = 2$. The proof will require the existence result of this chapter. More recently, this result, without any a-priori symmetry assumption, in the case of the seven dimensional quaternionic Heisenberg group was achieved in [95], and in any dimension in [96]. We will present the results of [96] and [95] in Chapter 6. The variational problem in this case and the associated Euler-Lagrange equation when $p = 2$ lead to the Yamabe type equation

$$\mathcal{L}u = -u^{2^*-1}, \tag{1.6}$$

where $\mathcal{L}u = \sum_{j=1}^{m} X_j^2 u$ and $\mathcal{L}$ is the corresponding sub-Laplacian, cf. (1.16).

1.2 Carnot groups

In this section we review some well known facts, see for example [82; 65; 21; 68; 132; 27], and set-up some of the notation we will use throughout the book.

A Carnot group is a connected and simply connected Lie group G with a Lie algebra $\mathfrak{g}$, which admits a stratification, i.e., $\mathfrak{g}$ is a direct sum of linear subspaces, $\mathfrak{g} = \bigoplus_{j=1}^{r} V_j$, with commutators satisfying the conditions $[V_1, V_j] = V_{j+1}$ for $1 \le j < r$, $[V_1, V_r] = \{0\}$, [65]. The number r is called the step of the Carnot group. It tells us how many commutators of the first layer are needed in order to generate the whole Lie algebra. Notice that the "last layer" V_r is the center of the Lie algebra, while the first layer V_1 generates the whole Lie algebra. Clearly, every Carnot group is trivially a nilpotent group.

As well known, see for example [[68], Proposition 1.2] and [[48], Theorem 1.2.1] for proofs, we have the following

Theorem 1.2.1. *If G is a connected and simply connected nilpotent Lie group with Lie algebra $\mathfrak{g}$, then:*

a) the exponential map $\exp : \mathfrak{g} \to G$ is an analytic diffeomorphism (with inverse denoted by log*);*

b) in exponential coordinates, i.e., if G is identified with $\mathfrak{g}$ via exp, *the group law is a polynomial map;*

c) the push-forward of the Lebesgue measure on $\mathfrak{g}$ is a bi-invariant Haar measure on G.

The Haar measure will be denoted by dH. According to part c), when we use exponential coordinates to compute integrals we need to use the Lebesgue measure on $\mathfrak{g}$. These results follow essentially from the Baker-Campbell-Hausdorff formula

$$\exp \xi \exp \eta = \exp\left(\xi + \eta + 1/2[\xi, \eta] + ...\right), \qquad \xi, \eta \in \mathfrak{g}, \qquad (1.7)$$

which for nilpotent groups reduces to a finite sum.

We shall use the exponential map to define analytic maps $\xi_i : G \to V_i, i = 1, ..., r$, through the equation $g = \exp \xi$, i.e., $\xi = \log g$, with

$$\xi(g) = \xi_1(g) + ... + \xi_r(g), \qquad \xi_i \in V_i.$$

Convention 1.2.2. We assume that on $\mathfrak{g}$ there is a scalar product $< ., . >$ with respect to which the V_j's are mutually orthogonal, i.e., $\mathfrak{g} = \overset{r}{\underset{j=1}{\oplus}} V_j$ is an orthogonal direct sum. We denote by $\| . \|$ the corresponding norm.

This fixed scalar product induces a left-invariant metric on G and a distance function $d_R(.,.)$ on G called the Riemannian distance. With the help of the exponential map the Riemannian distance on G is

$$d_R(g, h) = \|\log(h^{-1}g)\|, \qquad (1.8)$$

which is clearly left-invariant. However, this "Euclidean" distance does not preserve the homogeneous structure which we define next.

Every Carnot group is naturally equipped with a family of non-isotropic dilations defining the so-called *homogeneous structure*,

$$\delta_\lambda(g) = \exp \circ \Delta_\lambda \circ \exp^{-1}(g), \qquad g \in G, \qquad (1.9)$$

where $\exp : \mathfrak{g} \to G$ is the exponential map and $\Delta_\lambda : \mathfrak{g} \to \mathfrak{g}$ is defined by $\Delta_\lambda(\xi_1 + ... + \xi_r) = \lambda \xi_1 + ... + \lambda^r \xi_r$. The topological dimension of G is $N = \sum_{j=1}^{r} \dim V_j$, whereas the *homogeneous dimension* of G, attached to the automorphisms $\{\delta_\lambda\}_{\lambda > 0}$, is given by $Q = \sum_{j=1}^{r} j \dim V_j$. Recalling Theorem 1.2.1, we have

$$dH(\delta_\lambda(g)) = \lambda^Q dH(g), \qquad (1.10)$$

so that the number Q plays the role of a dimension with respect to the group dilations. Let Z be the infinitesimal generator of the one-parameter group of non-isotropic dilations $\{\delta_\lambda\}_{\lambda>0}$. Such vector field is characterized by the property that a function $u : G \to \mathbb{R}$ is homogeneous of degree s with respect to $\{\delta_\lambda\}_{\lambda>0}$, i.e., $u(\delta_\lambda(x)) = \lambda^s u(x)$ for every $x \in G$, if and only if $Zu = su$.

Finally, the above fixed norm $\|.\|$ on $\mathfrak{g}$ (arising from the fixed inner product on $\mathfrak{g}$) induces a *homogeneous norm (gauge)* $|\cdot|_{\mathfrak{g}}$ on $\mathfrak{g}$ and (via the exponential map) one on the group G in the following way. For $\xi \in \mathfrak{g}$, with $\xi = \xi_1 + ... + \xi_r$, $\xi_i \in V_i$, we let

$$|\xi|_{\mathfrak{g}} = \left(\sum_{i=1}^{r} \|\xi_i\|^{2r!/i} \right)^{\frac{1}{2r!}}, \tag{1.11}$$

and then define $|g|_G = |\xi|_{\mathfrak{g}}$ if $g = \exp \xi$. In particular, for a Carnot group of step two we have

$$|\xi|_{\mathfrak{g}} = \left(\|\xi_1\|^4 + \|\xi_2\|^2 \right)^{1/4}. \tag{1.12}$$

For simplicity of notation we shall use $|.|$ to denote both the norm on the group and on its Lie algebra with the meaning clear from the context. A homogeneous norm, see [110], [117] or [68], adapted to the fixed homogeneous structure is any function $|.| : G \to [0, +\infty)$ such that

$$|h| = 0 \text{ iff } h = e \ - \text{ the group identity,}$$
$$|h| = |h^{-1}|, \qquad |\delta_\lambda h| = \lambda|h|, \qquad |.| \in \mathcal{C}(G), \tag{1.13}$$

for any g_1, g_2, $h \in G$. The above defined gauge clearly satisfies these conditions. We shall call the balls defined with the help of (1.11) the *gauge balls* and denote them with $\Omega_r(g) \equiv \Omega(g,r)$, $r > 0$, thus

$$\Omega_r(g) \equiv \Omega(g,r) = \{h \in G : |g^{-1}h| < r\}. \tag{1.14}$$

The homogeneous norm is not unique and below we will see another widely used example defined with the help of the so called Carnot-Carathéodory distance. It is important to observe that the balls defined by a homogeneous norm are compact in the topology defined by the Riemannian distance on G. The proof of this fact uses only equation (1.13). Indeed, since the unit Riemannian sphere is compact and does not contain the neutral element it follows that $|.|$ achieves a positive minimum on the Riemannian sphere. The homogeneity of the gauge implies then that every gauge ball is a bounded and compact set. In particular we can integrate over gauge balls. Another consequence of this compactness is that in addition to the above properties we have for some $\gamma \geq 1$

$$|g_1 g_2| \leq \gamma (|g_1| + |g_2|)$$

for any g_1, g_2, $h \in \boldsymbol{G}$, see [110] . The compactness also implies that all homogeneous norms adapted to the fixed homogeneous structure are equivalent, i.e., if $|\,.\,|_1$ and $|\,.\,|_2$ are two such norms then there are constants M, $m > 0$ for which

$$m|g|_1 \leq |g|_2 \leq M|g|_1, \quad q \in \boldsymbol{G}.$$

A homogeneous norm on $\boldsymbol{G}$ can be used to define a homogeneous pseudo-distance on $\boldsymbol{G}$, which in the case of the fixed gauge is frequently called the gauge distance,

$$\rho(g_1, g_2) = |g_2^{-1} g_1|_{\boldsymbol{G}}, \quad \rho(\delta_\lambda(g_1), \delta_\lambda(g_2)) = \lambda \rho(g_1, g_2)$$
$$\rho(g_1, g_2) \leq \gamma (\rho(g_1, h) + \rho(h, g_2)), \quad g_1, g_2, h \in \boldsymbol{G}. \tag{1.15}$$

Notice that (1.11) has one additional property, which we shall not use, namely, it defines a smooth function outside the identity, i.e., $|\,.\,| \in \mathcal{C}(\boldsymbol{G}) \cap \mathcal{C}^\infty (\boldsymbol{G} \setminus \{e\})$. It is known that a smooth homogeneous norm which is actually a distance exists on every homogeneous group [87]. We shall encounter such a norm when we consider the gauge (2.12) on groups of Heisenberg type, see [54].

With $m = \dim(V_1)$, we fix a basis $X_1, \dots, X_m$ of V_1 and continue to denote by the same letters the corresponding left-invariant vector fields. The *sub-Laplacian* associated with X is the second-order partial differential operator on $\boldsymbol{G}$ given by

$$\mathcal{L} = -\sum_{j=1}^m X_j^* X_j = \sum_{j=1}^m X_j^2, \tag{1.16}$$

noting that in a Carnot group one has $X_j^* = -X_j$ since the Haar measure is bi-invariant under translations.

Since the left invariant vector fields $\{X_1, \dots, X_m\}$ on $\boldsymbol{G}$ satisfy the Hörmander finite rank condition we can define also the associated *Carnot-Carathéodory* distance on $\boldsymbol{G}$. A piecewise smooth $\gamma(t)$ is called a horizontal curve when $\gamma'(t)$ belongs to the span of $X_1(\gamma(t)), \dots, X_m(\gamma(t))$, the so called horizontal space. Given $g, h \in \boldsymbol{G}$ by the Chow-Rashevsky theorem there is a horizontal curve between them, see [46], [141]. Therefore

$$d(g, h) = \inf \{|\gamma| : \gamma - \text{horizontal curve between } g \text{ and } h\}, \tag{1.17}$$

defines a distance on $\boldsymbol{G}$, which is the Carnot-Carathéodory distance between g and h, see also [130]. Here $|\gamma|$ denotes the length of γ with respect to the fixed left invariant Riemannian metric on $\boldsymbol{G}$. Note that only the metric on the horizontal space is used to calculate the length of a horizontal curve. Taking into account that we fixed a left invariant metric which is homogeneous of order one on the horizontal space it follows that for every $f, g, h \in \boldsymbol{G}$ and for any $\lambda > 0$

$$d(gf, gh) = d(f, h), \qquad d(\delta_\lambda(g), \delta_\lambda(h)) = \lambda\, d(g, h). \tag{1.18}$$

In particular, $g \mapsto d(e, g)$ is a homogeneous norm on the considered Carnot group. It is straightforward to estimate the Riemannian distance by the Carnot-Carathéodory distance,

$$d_R(g, h) \leq d(g, h), \tag{1.19}$$

which, in particular, shows that $d(g, h) = 0$ iff $g = h$. This can be used to see that $d(g, h)$ is a distance function since the other properties follow from the definition. The estimate in the other direction can be found for example in [160], where a more general situation is considered. In our case we obtain that for every ball $B(g_o, R)$ there exists a constant $C = C(\boldsymbol{G}, R)$ such that if $g, h \in B(g_o, R)$ we have

$$d(g, h) \leq C\, d_R(g, h)^{1/r}. \tag{1.20}$$

The pseudo-distance (1.15) is equivalent to the Carnot-Carathéodory distance $d(\cdot, \cdot)$ generated by the system $\{X_1, ..., X_m\}$, i.e., there exists a constant $C = C(\boldsymbol{G}) > 0$ such that

$$C\, \rho(g, h) \leq d(g, h) \leq C^{-1}\, \rho(g, h), \qquad g, h \in \boldsymbol{G}, \tag{1.21}$$

taking into account the equivalence of all homogeneous norms. In fact the inequalities (1.19) and (1.20) can be derived for the distance defined by any homogeneous norm and the Riemannian distance, see [65].

We will almost exclusively work with the Carnot-Carathéodory distance d, except in few situations where we will find more convenient to use the gauge distance (1.15). In general, we shall work with the Carnot-Carathéodory balls which are defined in the obvious way,

$$B_r(x) \equiv B(x, R) = \{y \in \boldsymbol{G} \mid d(x, y) < r\}, \qquad r > 0.$$

By left-translation and dilation it is easy to see that the Haar measure of $B(x, r)$ is proportional to r^Q, where Q is the homogeneous dimension of $\boldsymbol{G}$.

1.3   Sobolev spaces and their weak topologies

Let $1 \leq p < \infty$ and Ω be an open set in G. We define the space $\mathcal{S}^{1,p}(\Omega)$ of all functions $u \in L^p(\Omega)$ having a distributional horizontal gradient $Xu = (X_1 u, ..., X_m u) \in L^p(\Omega)$. The space $\mathcal{S}^{1,p}(\Omega)$ will be endowed with the norm

$$||u||_{\mathcal{S}^{1,p}(\Omega)} = ||u||_{L^p(\Omega)} + ||Xu||_{L^p(\Omega)},$$

which turns it into a Banach space using the corresponding results in Euclidean space. Here $||Xu||_{L^p(\Omega)} = \left[\int_\Omega |Xu|^p \, dH \right]^{1/p}$. It is important to note a version of the Rellich-Kondrachov compact embedding valid in the sub-elliptic setting which was proven in [74], see also [69] for a proof of part b).

Theorem 1.3.1.

a) If Ω denotes a bounded X-PS domain (Poincaré-Sobolev domain) in a Carnot-Carathéodory space and $1 \leq p < Q$, then the embedding

$$\mathcal{S}^{1,p}(\Omega) \subset L^q(\Omega)$$

is compact provided that $1 \leq q < p^*$, where Q is the homogeneous dimension of the group G and $p^* = pQ/(Q-p)$ is the Sobolev exponent relative to p.

b) The Carnot-Carathéodory balls are X-PS domains.

We also define the space $\overset{o}{\mathcal{D}}{}^{1,p}(\Omega)$, $1 \leq p < Q$, by taking the completion of the space of all smooth functions with compact support $\mathcal{C}_o^\infty(\Omega)$ with respect to the norm

$$||u||_{\overset{o}{\mathcal{D}}{}^{1,p}(\Omega)} = ||Xu||_{L^p(\Omega)} = \left[\int_\Omega |Xu|^p \, dH \right]^{1/p}.$$

We shall consider $\overset{o}{\mathcal{D}}{}^{1,p}(\Omega)$ equipped with the norm $||u||_{\overset{o}{\mathcal{D}}{}^{1,p}(\Omega)}$. Notice that because of the Folland-Stein's inequality every element of $\overset{o}{\mathcal{D}}{}^{1,p}(\Omega)$ is in fact a function $u \in L^{p^*}(\Omega)$. As usual, one can see that $u \in \overset{o}{\mathcal{D}}{}^{1,p}(\Omega)$ implies that $|u| \in \overset{o}{\mathcal{D}}{}^{1,p}(\Omega)$.

The space $\overset{o}{\mathcal{D}}{}^{1,p}(\Omega)$ and, in particular, $\overset{o}{\mathcal{D}}{}^{1,p}(G)$ are the most important Sobolev spaces in this book. Both are easily seen to be Banach spaces - see the next Proposition. Given a Banach space $\mathcal{B}$ we consider the dual space $\mathcal{B}^*$ of all continuous linear functionals on $\mathcal{B}$ equipped with the norm $||l|| = \sup_{x \in \mathcal{B}, ||x|| \leq 1} |l(x)|$, $\lambda \in \mathcal{B}^*$. As well known $\mathcal{B}^*$ is also a Banach space with this topology, see for example [145] for details. The space $\mathcal{B}$ is called reflexive if its second dual $\mathcal{B}^{**} = (\mathcal{B}^*)^*$ equals $\mathcal{B}$. Of course, $\mathcal{B} \subset \mathcal{B}^{**}$ is an isometric injection [[175], p.113, Theorem 2], which might not be a surjection. The latter happens exactly when $\mathcal{B}$ is reflexive. It is known that L^p spaces are reflexive for $p > 1$ and L^1 is not reflexive. In fact for $1 \leq p < \infty$ we have $(L^p)^* = L^{p'}, \frac{1}{p} + \frac{1}{p'} = 1$. In particular, $\left(L^1(G)\right)^* = L^\infty(G)$, where the latter space is equipped with the norm

$$\|u\|_{L^\infty(G)} \equiv \operatorname*{esssup}_G |u| \overset{def}{=} \inf\{M : |u(g)| \leq M \text{ for } dH \text{ a.e. } g \in G\}.$$

$$(1.22)$$

On the other hand, the dual of $L^\infty(G)$ is the space of all absolutely continuous (w.r.t. the Haar measure), finitely additive set functions of bounded total variation on G, see [[175], Chapter IV.9]. The dual Banach space of $\overset{o}{\mathcal{D}}{}^{1,p}(\Omega)$ will be denoted by $\mathcal{D}^{-1,p'}(\Omega)$ and is identified in the next Proposition. Notice that the density of $\mathcal{C}_o^\infty(\Omega)$ in $\overset{o}{\mathcal{D}}{}^{1,p}(\Omega)$ implies that a continuous linear functional on the latter space is determined by its action on $\mathcal{C}_o^\infty(\Omega)$.

Proposition 1.3.2. *Let Ω be an open set in a Carnot group whose first layer is of dimension m. Let $1 \leq p < \infty$ and p' be the Hölder conjugate to $p, \frac{1}{p'} + \frac{1}{p} = 1$.*

a) The space $\overset{o}{\mathcal{D}}{}^{1,p}(\Omega)$ is a separable Banach space, which is reflexive if $1 < p < \infty$.

b) The dual space $\overset{o}{\mathcal{D}}{}^{1,p}(\Omega)$ is isometric to the Banach space of distributions

$$\mathcal{D}^{-1,p'}(\Omega) = \{T \in \mathcal{D}'(\Omega) : T = \sum_{j=1}^m X_j f_j, \ f_j \in L^{p'}(\Omega)\}, \qquad (1.23)$$

with the pairing between a function $u \in \mathcal{C}_o^\infty(\Omega)$ and a distribution $T_f = \sum_{j=1}^m X_j f_j$ given in the usual fashion

$$T(u) = -\int_\Omega \sum_{j=1}^m f_j (X_j u) \, dH. \qquad (1.24)$$

The corresponding norm is

$$\|T\|_{\mathcal{D}^{-1,p'}(\Omega)} = \inf\|\vec{f}\|_{L^{p'}(\Omega)}, \qquad (1.25)$$

where the $\inf$ is taken over all $\vec{f} = (f_1, \ldots, f_m) \in L^{p'}(\Omega : \mathbb{R}^m)$ for which the representation $T = \sum_{j=1}^m X_j f_j$ holds true and $\|\vec{f}\|_{L^{p'}(\Omega)} =$

$$\left[\int_\Omega \left(\sum_{j=1}^m |f_j|^2 \right)^{p'/2} dH \right]^{1/p'} \quad \text{for } p' < \infty, \text{ and the corresponding sup norm}$$

when $p' = \infty$. Furthermore, $\mathcal{D}^{-1,p'}(\Omega)$ is a separable Banach space, which is reflexive if $1 < p < \infty$.

Proof. a) By definition the map $X : \overset{o}{\mathcal{D}}{}^{1,p}(\Omega) \to L^p(\Omega : R^m)$ given by $X(u) = (X_1 u, \ldots, X_m u)$ is an isometry. If $\{u_n\}$ is a Cauchy sequence in $\overset{o}{\mathcal{D}}{}^{1,p}(\Omega)$ it follows from the completeness of $L^q(\Omega)$, $1 \le q$, for every $j = 1, \ldots, m$ we have $X_j u_n \to f_j$ in $L^p(\Omega)$, while Folland and Stein's embedding theorem implies $u_n \to u$ in $L^{p^*}(\Omega)$. For any function $\varphi \in C_o^\infty(\Omega)$ we have then

$$\int_\Omega f_j \varphi \, dH = \lim_{n \to \infty} \int_\Omega (X_j u_n) \varphi \, dH$$

$$= -\lim_{n \to \infty} \int_\Omega u_n (X_j \varphi) \, dH = -\int_\Omega u(X_j \varphi) \, dH,$$

taking into account $X_j^* = -X_j$, hence $X_j u = f_j$, i.e., $u \in \overset{o}{\mathcal{D}}{}^{1,p}(\Omega)$.

In particular, $\overset{o}{\mathcal{D}}{}^{1,p}(\Omega)$ embeds isometrically in $L^p(\Omega : R^m)$ as a subspace, i.e., the image is a closed subspace. Since $L^p(\Omega : R^m)$ is a separable Banach space, which is reflexive if $1 < p < \infty$, the same remains true for every closed subspace.

b) The proofs are analogous to the Euclidean case, see for example [1] and [128]. The fact that every distribution of the described type defines a linear bounded functional on $\overset{o}{\mathcal{D}}{}^{1,p}(\Omega)$ follows from the Cauchy-Schwarz and Hölder's inequalities,

$$|T(u)| = \left| \int_\Omega \sum_{j=1}^m f_j (X_j u) \, dH \right| \le \int_\Omega |\vec{f}| \, |Xu| \, dH \le \|\vec{f}\|_{L^{p'}(\Omega)} \|Xu\|_{L^p(\Omega)},$$

where $|\vec{f}| = \left(\sum_{j=1}^m |f_j|^2 \right)^{1/2}$.

The other direction, i.e., that every linear bounded functional T on $\overset{o}{\mathcal{D}}{}^{1,p}(\Omega)$ can be written in the form $T = -\sum_{j=1}^m X_j f_j$, $f_j \in L^{p'}(\Omega)$ can be seen as follows. Let $\mathfrak{X}$ be the image of $\overset{o}{\mathcal{D}}{}^{1,p}(\Omega)$ under the isometry X. The linear functional T' on $\mathfrak{X}$ defined by $T'(Xu) = T(u)$ is clearly bounded and $\|T'\| = \|T\|$. By the Hahn-Banach theorem there exists a norm preserving extension of T' to a bounded linear functional on $L^p(\Omega : R^m)$, which by the Riesz representation theorem is a unique element $\vec{f} \in L^{p'}(\Omega : R^m)$ and has the form $T'(\vec{g}) = \sum_{j=1}^{m} \int_\Omega f_j g_j \, dH$. Thus

$$T(u) = T'(Xu) = \sum_{j=1}^{m} \int_\Omega f_j(X_j u) \, dH,$$

which shows $T = -\sum_{j=1}^{m} X_j f_j$ and $\|T\| = \|T'\| = \|\vec{f}\|_{L^{p'}(\Omega)}$. Let us observe that any other $\vec{f}^* \in L^{p'}(\Omega : R^m)$ for which $T = -\sum_{j=1}^{m} X_j f_j^*$ corresponds to another extension of T' from $\mathfrak{X}$ to $L^p(\Omega : R^m)$, which in general will increase its norm. In fact, the vector $\vec{f}$ we found above corresponds to the unique norm preserving extension at least when $1 < p < \infty$. Finally, the completeness of $\mathcal{D}^{-1,p'}(\Omega)$ follows from the isometry $\|T\| = \|T'\| = \|\vec{f}\|_{L^{p'}(\Omega)}$.

The last claim of the Proposition follows from the fact that $\mathfrak{X}$ is a subspace of the Banach space $L^p(\Omega : \mathbb{R}^m)$ which has the listed properties. $\square$

Given a Banach space $\mathcal{B}$ with dual $\mathcal{B}^*$ we define the weak topology, denoted by $\rightharpoonup$ on $\mathcal{B}$, see for example [145], as the *weakest topology* on $\mathcal{B}$ for which all elements of $\mathcal{B}^*$ are continuous functionals on $\mathcal{B}$. In particular, $x_n \in \mathcal{B} \rightharpoonup x \in \mathcal{B}$ if $l(x_n) \to l(x)$ for every $l \in \mathcal{B}^*$.

We shall also make use of the weak-$*$ topology on $\mathcal{B}^*$ that is the weakest topology on $\mathcal{B}^*$ in which for all $x \in \mathcal{B}$ the functionals $l \in \mathcal{B}^* \mapsto l(x)$ are continuous. Thus, a sequence $l_n \in \mathcal{B}^*$ converges to $l \in \mathcal{B}^*$ in weak-$*$ sense if $l_n(x) \to l(x)$ for every $x \in \mathcal{B}$. Notice that this implies that the weak-$*$ convergence on $\mathcal{B}^*$ is even weaker than the weak convergence on $\mathcal{B}^*$ since the topology of the former is defined by using only the elements of $\mathcal{B} \subset \mathcal{B}^{**}$, while the latter topology is determined by all elements of $\mathcal{B}^{**}$. In particular, for a reflexive space $\mathcal{B}$ we can regard $\mathcal{B}$ as the dual of $\mathcal{X} = \mathcal{B}^*$, $\mathcal{B} = \mathcal{X}^*$, and thus the weak-$*$ topology of $\mathcal{B}$ is determined by the elements of X, while the weak topology of $\mathcal{B}$ is determined by the elements of $\mathcal{B}^*$. Since $\mathcal{B}^* = \mathcal{X}$, the two naturally defined weak topologies on a reflexive Banach space coincide. On the other hand, if we consider the non reflexive space $L^1(G)$ we have that its dual is $L^\infty(G)$, and thus for a sequence $u_n \in L^\infty(G)$ we have $u_n \rightharpoonup f$ in the weak-$*$ topology of $L^\infty(G)$ when for each $v \in L^1(G)$ we have

$$\int_G u_n v \, dH \to \int_G f v \, dH.$$

A basic fact in the theory is the Banach-Alaoglu theorem [145]. It says that in the dual space $\mathcal{B}^*$ (of a Banach space!) equipped with the weak-$*$ topology the closed unit ball is compact. Furthermore, if a sequence

$l_n \in \mathcal{B}^* \rightharpoonup l \in \mathcal{B}^*$ in the weak-$*$ topology, then $\{l_n\}$ is a bounded sequence and the norm is sequentially lower semi-continuous [[175], p.125, Theorem 9],

$$\|l\| \leq \liminf \|l_n\|.$$

For our goals, we are actually interested in sequential compactness, i.e., in the possibility of extracting a convergent subsequence from a bounded sequence. Very often it is useful to use a weaker version of the Banach-Alaoglu's theorem, namely Helly's theorem, which is the fact that if $\mathcal{B}$ is a separable Banach space then the closed unit ball of $\mathcal{B}^*$ is weak-$*$ sequentially compact. With the help of this general results, Proposition 1.3.2 and the Folland-Stein embedding theorem we obtain the following weak compactness result.

Proposition 1.3.3. *Let Ω be an open subset of G. If $\{u_n\}$ is a bounded sequence in $\overset{o}{\mathcal{D}}{}^{1,p}(\Omega)$, $p > 1$, then it has a subsequence v_n which converges to a function v in:*
i) weak sense in $\overset{o}{\mathcal{D}}{}^{1,p}(\Omega)$, ii) weak sense in $L^{p^*}(\Omega)$.
Furthermore, by the sequentially lower semi-continuity of the norms we have

$$\|u\|_{L^{p^*}(\Omega)} \leq \liminf \|u_n\|_{L^{p^*}(\Omega)}, \qquad \|u\|_{\overset{o}{\mathcal{D}}{}^{1,p}(\Omega)} \leq \liminf \|u_n\|_{\overset{o}{\mathcal{D}}{}^{1,p}(\Omega)}.$$

For our needs we shall also need to consider the weak-$*$ topology in the space $\mathcal{M}(G)$ of all bounded regular Borel measures on G. Recall that $\mathcal{M}(G)$ can be considered as the dual space of the Banach space $\mathcal{C}_\infty(G)$ of all continuous functions on G which vanish at infinity. A continuous function ϕ on G is said to vanish at infinity if for every $\epsilon > $ there is a compact $K \subset G$ such that $|\phi(g)| < \epsilon$, $g \in G \setminus K$. Alternatively, since G is a locally compact Hausdorff space, starting with the space of all continuous functions on G with compact support $\mathcal{C}_o(G)$ we have that $\mathcal{C}_\infty(G)$ is the completion of the space $\mathcal{C}_o(G)$ in the $L^\infty(G)$ norm. The proofs of these facts can be found in [[146], Chapter 3]. By the Riesz representation theorem [[146], Chapter 6], the space $\mathcal{M}(G)$ is the dual of $\mathcal{C}_\infty(G)$ and in particular for a sequence of measures $d\nu_n$ (bounded regular Borel measures on G) to converge to a measure $d\nu$ in the weak-$*$ topology of $\mathcal{M}(G)$ it is necessary and sufficient that for every $\phi \in \mathcal{C}_o(G)$ we have

$$\int_G \phi \, d\nu_n \to \int_G \phi \, d\nu.$$

The following results will play an important role in the study of the variational problem (1.3).

Proposition 1.3.4. *If $\{u_n\}$ is a bounded sequence in $\overset{o}{\mathcal{D}}{}^{1,p}(\boldsymbol{G})$, $p > 1$, then it has a subsequence v_n for which in addition to the convergence of Proposition 1.3.3 the following convergence to the function v take place:*

i) $d\mu_n = |Xv_n|^p dH \rightharpoonup d\mu$ and $d\nu_n = |v_n|^{p^*} dH \rightharpoonup d\nu$ in the weak-* topology of $\mathcal{M}(\boldsymbol{G})$ for some $d\mu$, $d\nu \in \mathcal{M}(\boldsymbol{G})$;

ii) a.e. point-wise on $\boldsymbol{G}$;

Proof. By the Folland-Stein embedding theorem both $|Xu_n|^p$ and $|u_n|^{p^*}$ are bounded sequences in $L^1(\boldsymbol{G})$, hence bounded sequences in $\mathcal{M}(\boldsymbol{G})$ considered with the weak-* topology, as for $f \in L^1(\boldsymbol{G})$ we have $|\int_{\boldsymbol{G}} \phi f\, dH| \leq \|\phi\|_{L^\infty(\boldsymbol{G})} \|f\|_{L^1(\boldsymbol{G})}$, $\phi \in \mathcal{C}_\infty(\boldsymbol{G})$. The claim in i) follows then from the Banach-Alaoglou theorem.

The a.e. convergence follows from the Rellich-Kondrachov Theorem 1.3.1 applied to an increasing sequence of Carnot-Carathéodory balls B_{R_k} centered at the identity $e \in \boldsymbol{G}$ with radius $R_k \to \infty$ so that $B_{R_k} \subset B_{R_{k+1}} \subset \boldsymbol{G}$, $\cup B_{R_k} = \boldsymbol{G}$. Notice that $\overset{o}{\mathcal{D}}{}^{1,p}(\Omega) \subset L^p(\Omega)$ being compact embedding when Ω is an open bounded X-PS domain implies that weakly convergent sequences in $\overset{o}{\mathcal{D}}{}^{1,p}(\Omega)$ are convergent in $L^p(\Omega)$. $\qquad\square$

Let us observe that $d\nu \neq |v|^{p^*} dH$ in general! This is because the measure $|v_n - v|^{p^*} dH$ can concentrate on a set of measure zero. Nevertheless, even when v_n fails to converge to v in $L^{p^*}(\Omega)$ we have the following result of Brézis-Lieb [32], frequently called the Brézis-Lieb lemma, which will be used in the proof of Lemma 1.4.5.

Theorem 1.3.5. *Let $0 < q < \infty$. If $u_n \to u$ a.e. and $\|u_n\|_{L^q(\boldsymbol{G})} \leq C < \infty$ for all $n \in \mathbb{N}$, then the limit $\lim_{n \to \infty}\left\{\|u_n\|_{L^q(\boldsymbol{G})}^q - \|u_n - u\|_{L^q(\boldsymbol{G})}^q\right\}$ exists and we have the equality*

$$\lim_{n \to \infty}\left\{\|u_n\|_{L^q(\boldsymbol{G})}^q - \|u_n - v\|_{L^q(\boldsymbol{G})}^q\right\} = \|u\|_{L^q(\boldsymbol{G})}^q.$$

In particular we obtain

Corollary 1.3.6. *Under the conditions of the above Theorem 1.3.5 we have*

$$|u_n|^q\, dH - |u|^q\, dH = |u_n - u|^q\, dH - o(1), \qquad (1.26)$$

where $o(1) \rightharpoonup 0$ weak- in $\mathcal{M}(\boldsymbol{G})$.*

Proof. For $\phi \in \mathcal{C}_\infty(G)$ consider the sequence ϕu_n, which has the properties $\phi u_n \to \phi u$ a.e. and $\sup_n \|\phi u_n\|_{L^q(G)} < \infty$. By the Brézis-Lieb lemma we have

$$\int_G |\phi|^q |u_n|^q \, dH - \int_G |\phi|^q |u_n - u|^q \, dH \to \int_G |\phi|^q |u|^q \, dH, \quad n \to \infty.$$

This implies equation (1.26) as every function in $\mathcal{C}_\infty(G)$ can be written as the difference of its positive and negative parts, which are also in $\mathcal{C}_\infty(G)$. $\qquad\square$

1.4 The best constant in the Folland-Stein inequality

In this section we turn to the existence of minimizers of the problem (1.3), i.e., we shall prove the main result on existence of global minimizers, [166]. Before stating the main Theorem 1.4.2 we define the relevant spaces of functions. As before $X = \{X_1, \ldots, X_m\}$ is the fixed left-invariant basis of the first layer V_1 generating the whole Lie algebra and also, with abuse of notation, the corresponding system of sections on G. Correspondingly, for a function u on G we shall use the notation $|Xu| = \left(\sum_{j=1}^{m} (X_j u)^2 \right)^{1/2}$ for the length of the horizontal gradient. Two crucial aspects of equation (1.2) and variational problem (1.3) are their invariance with respect to the group translations and dilations, which are the cause of such concentration phenomenon. The invariance with respect to translations is obvious, since the vector fields X_j are left-invariant. The invariance with respect to scaling must be suitably interpreted and follows from the observation that

$$\mathcal{L}_p(u \circ \delta_\lambda) = \lambda^p \, \delta_\lambda \circ \mathcal{L}_p u. \tag{1.27}$$

If we then define, for a solution u of (1.2) and for $\lambda > 0$, the rescaled function $u_\lambda = \lambda^\alpha u \circ \delta_\lambda$, it is clear that u_λ satisfies (1.2) if and only if $\alpha = Q/p^* = (Q - p)/p$. These considerations suggest the introduction of two new functions. For a function $u \in C_o^\infty(G)$ we let

$$\tau_h u \overset{def}{=} u \circ \tau_h, \qquad h \in G, \tag{1.28}$$

where $\tau_h : G \to G$ is the operator of left-translation $\tau_h(g) = hg$, and also

$$u_\lambda \equiv \lambda^{Q/p^*} \delta_\lambda u \overset{def}{=} \lambda^{Q/p^*} u \circ \delta_\lambda, \qquad \lambda > 0. \tag{1.29}$$

It is easy to see that the norms in the Folland-Stein inequality and the functionals in the variational problem (1.3) are invariant under the translations

(1.28) and the rescaling (1.29). Only the second part requires a small computation since the measure dH is bi-invariant with respect to translations. Now,

$$\|\delta_\lambda u\|^{p^*}_{L^{p^*}(G)} = \int_G |u(\delta_\lambda g)|^{p^*} dH(g)$$

$$= \int_G |u(g)|^{p^*} \lambda^{-Q} dH(g) = \lambda^{-Q} \|u\|^{p^*}_{L^{p^*}(G)}.$$

hence $\|u_\lambda\|_{L^{p^*}(G)} = \|u\|_{L^{p^*}(G)}$. Similarly

$$\|\delta_\lambda u\|^p_{\overset{o}{\mathcal{D}}{}^{1,p}(G)} = \|\lambda \delta_\lambda X u\|^p_{L^p(G)}$$

$$= \lambda^p \|\delta_\lambda X u\|^p_{L^p(G)} = \lambda^{p-Q} \|u\|^p_{\overset{o}{\mathcal{D}}{}^{1,p}(G)}.$$

Taking into account $p^* = \frac{pQ}{Q-p}$ we obtain the claimed invariance $\|Xu_\lambda\|_{\overset{o}{\mathcal{D}}{}^{1,p}(G)} = \|Xu\|_{\overset{o}{\mathcal{D}}{}^{1,p}(G)}$.

At this point we are ready to turn to the main result of this section, Theorem 1.4.2, which is based on an adaptation of the concentration-compactness principle of P.L. Lions to the case of a Carnot group G with its homogeneous structure and Carnot-Carathéodory distance. An important tool in the analysis is the concentration function of a measure, which is given in the next definition.

Definition 1.4.1.

a) For a non-negative measure $d\nu$ on G define the concentration function Q on $[0, \infty)$ by

$$Q(r) \overset{def}{=} \sup_{g \in G} \left(\int_{B_r(g)} d\nu \right). \qquad (1.30)$$

b) For a function $f \in L^{p^*}_{loc}(G)$ on G we will call concentration function of f the concentration function of the measure $|f|^{p^*} dH$.

Following Lions' work, the crucial ingredients in the solution of the variational problem of Theorem 1.4.2 are Lemmas 1.4.3 and 1.4.5, which follow its proof.

Theorem 1.4.2.

a) Let G be a Carnot group. Every minimizing sequence $u_n \in \overset{o}{\mathcal{D}}{}^{1,p}(G)$ of the variational problem (1.3),

$$\int_G |Xu_n|^p dH \to I, \qquad \int_G |u_n|^{p^*} dH = 1, \qquad (1.31)$$

has a convergent subsequence in $\overset{o}{\mathcal{D}}{}^{1,p}(G)$, after possibly translating and dilating each of its elements using (1.28) and (1.29).

b) The infimum in (1.3) is achieved by a non-negative function $u \in \overset{o}{\mathcal{D}}{}^{1,p}(G)$ which is a (weak) solution of the equation

$$\mathcal{L}_p u = - u^{p^*-1}. \tag{1.32}$$

Proof. The claim in part b) of the Theorem follows trivially from part a) together with the fact that for any $u \in \overset{o}{\mathcal{D}}{}^{1,p}(\Omega)$ we have that $|u| \in \overset{o}{\mathcal{D}}{}^{1,p}(\Omega)$ and $|Xu| = |X|u||$ a.e.. Thus it is enough to show the existence of a minimizer.

Consider the variational problem (1.3), $p > 1$, and a minimizing sequence $u_n \in C_o^\infty(G)$ as in (1.31). By the weak compactness, Proposition 1.3.3, we can assume that $u_n \rightharpoonup u$ in $\overset{o}{\mathcal{D}}{}^{1,p}(G)$ for some $u \in \overset{o}{\mathcal{D}}{}^{1,p}(G)$ and also weakly in $L^{p^*}(G)$. The sequentially lower semi-continuity of the norms shows that

$$\|u\|_{L^{p^*}(G)} \leq \liminf \|u_n\|_{L^{p^*}(G)} = 1,$$

and

$$\|u\|_{\overset{o}{\mathcal{D}}{}^{1,p}(G)} \leq \liminf \|u_n\|_{\overset{o}{\mathcal{D}}{}^{1,p}(G)} = I^{1/p}.$$

Thus, it is enough to prove that $\int_G |u|^{p^*} = 1$, since by the above and the Folland-Stein inequality (1.1), noting that $I = (1/S_p)^p$ (S_p is the optimal constant in (1.1)), we have

$$I \geq \int_G |Xu|^p \geq I\left(\int_G |u|^{p^*}\right)^{p/p^*} = I, \quad \text{when} \quad \int_G |u|^{p^*} = 1, \tag{1.33}$$

which would give that u is a minimizer of problem (1.3). In other words, we reduce to showing that $u_n \to u$ in $L^{p^*}(G)$ as for $1 < p < \infty$ weak convergence and convergence of the norms to the norm of the weak limit imply strong convergence.

Due to the translation and dilation invariance, all the above properties hold if we replace the sequence $\{u_n\}$ with any translated and rescaled sequence $\{v_n\}$ with corresponding weak limit denoted by v. We will consider the following measures,

$$d\nu_n \overset{def}{=} |v_n|^{p^*} dH \quad \text{and} \quad d\mu_n \overset{def}{=} |Xv_n|^p dH, \tag{1.34}$$

where v_n is a suitable translation and dilation of u_n that is to be defined in a moment. From the weak-$*$ compactness of the unit ball without loss of generality we can assume $d\nu_n \rightharpoonup d\nu$ and $d\mu_n \rightharpoonup d\mu$ in the weak-$*$ topology of all bounded, nonnegative measures, cf. Proposition 1.3.4. In addition, again by Proposition 1.3.4, we can also assume $v_n \to v$ a.e. with respect to the Haar measure.

The desired convergence i.e. the fact that $\int_G |v|^{p^*} = 1$ will be obtained by applying the concentration compactness principle exactly as in [125] (see also [155]). We shall see that ν is a probability measure and also $d\nu = \int_G |v|^{p^*}$.

Let $\hat{Q}_n(r)$ be the concentration function of u_n, i.e.,

$$\hat{Q}_n(r) \stackrel{def}{=} \sup_{h \in G} \left(\int_{B_r(h)} |u_n|^{p^*} dH \right). \tag{1.35}$$

Clearly, $\hat{Q}_n(0) = 0$, $\lim_{r \to \infty} Q_n(r) = 1$ and $\hat{Q}_n$ is a continuous non-decreasing function. Therefore, for every n we can find an $r_n > 0$ such that

$$\hat{Q}_n(r_n) = 1/2. \tag{1.36}$$

Since the integral in (1.35) is absolutely continuous, it defines a continuous function of h, which tends to zero when $d(h, e) \to \infty$ as $u_n \in L^{p^*}(G)$. Consequently, the sup is achieved, i.e., for every n there exist a $h_n \in G$, such that

$$\hat{Q}_n(r_n) = \int_{B_{r_n}(h_n)} |u_n(g)|^{p^*} dH(g). \tag{1.37}$$

The concentration functions, Q_n, of the dilated and translated sequence

$$v_n \stackrel{def}{=} r_n^{Q/p^*} u \circ \tau_{h_n^{-1}} \circ \delta_{r_n} = \left(\tau_{h_n^{-1}} u \right)_{r_n} \tag{1.38}$$

satisfy

$$Q_n(1) = \int_{B_1(e)} d\nu_n \qquad \text{and} \qquad Q_n(1) = \frac{1}{2}, \tag{1.39}$$

which follows easily from (1.28), (1.29), (1.18) and (1.10).

At this point we are ready to apply the key lemmas and we proceed the proof by considering the function v_n. Notice that the vanishing case in Lemma 1.4.3 is ruled out by the normalization $Q_n(1) = 1/2$. Following Lions, let us embed our variational problem in the family

$$I_\lambda \stackrel{def}{=} \inf \left\{ \int_G |Xu|^p \; : \; u \in C_o^\infty(G), \; \int_G |u|^{p^*} = \lambda \right\}. \tag{1.40}$$

Since $I_\lambda = \lambda^{p/p^*} I$, where I was defined in (1.31), we see that I_λ is strictly sub-additive i.e.

$$I_1 < I_\alpha + I_{1-\alpha}, \text{ for every } 0 < \alpha < 1. \tag{1.41}$$

Assume, first, the compactness case of Lemma 1.4.3 holds when applied to the sequence $d\nu_n$, which was defined in (1.34). Let g_n be a sequence of points given in the compactness case of Lemma 1.4.3. For $\epsilon > 0$ choose $R = R(\epsilon)$ such that

$$\int_{B_R(g_n)} d\nu_n \geq 1 - \epsilon, \text{ for every } n. \tag{1.42}$$

If $\epsilon < 1/2$ then $\int_{B_R(g_n)} d\nu_n > 1/2$ and since by construction $\int_{B_1(e)} d\nu_n = 1/2$ while $\int_G d\nu_n = 1$ we see that $B_1(e)$ and $B_R(g_n)$ have a non-empty intersection and thus by the triangle inequality $B_R(g_n) \subset B_{2R+1}(e)$. This implies

$$\int_{B_{2R+1}(e)} d\nu_n \geq \int_{B_R(g_n)} d\nu_n \geq 1 - \epsilon, \text{ for every } n, \tag{1.43}$$

and therefore the conclusions in the compactness case of Lemma 1.4.3 holds with $g_n \equiv e$ for every n. By taking $\epsilon \to 0$ we see that

$$\int_G d\nu = 1. \tag{1.44}$$

If we now look at the sequence $\{d\mu_n\}$, we have $d\mu_n \rightharpoonup d\mu$ and $\int_G d\mu_n \to I$, and thus $\int_G d\mu \leq I$. On the other hand, Lemma 1.4.5 gives

$$d\nu_n \rightharpoonup d\nu = |v|^{p^*} + \sum_j \nu_j \delta_{g_j}$$
$$d\mu_n \rightharpoonup d\mu \geq |Xv|^p dH + \sum_j \mu_j \delta_{g_j} \tag{1.45}$$

for certain $\nu_j, \mu_j \geq 0$ satisfying

$$I\nu_j^{p/p^*} \leq \mu_j. \tag{1.46}$$

We shall prove that all ν_j's are zero and thus $\int_G |v|^{p^*} dH = 1$. Let $\alpha \overset{def}{=} \int_G |v|^{p^*} dH < 1$. Since $\int_G d\nu = 1$ we have $\sum \nu_j = 1 - \alpha$. From $\int_G d\mu \leq I$ we have $\int_G |Xv|^p dH \leq I - \sum \mu_j$. Now (1.46) gives

$$I = I_1 \geq \int_G |Xv|^p dH + \sum \mu_j \geq I_\alpha + \sum I\nu_j^{p/p^*} \geq I_\alpha + \sum I_{\nu_j}.$$

From the strict sub-additivity (1.41) of I_λ we conclude that exactly one of the numbers α and ν_j is different from zero. We claim that $\alpha = 1$ (and thus all ν_j's are zero). Suppose that there is a $\nu_j = 1$ and $d\nu = \delta_{g_j}$. From the normalization $Q_n(1) = 1/2$ and hence

$$1/2 \geq \int_{B_1(g_j)} |v_n|^{p^*} dH \to \int_{B_1(g_j)} d\nu = 1, \qquad (1.47)$$

which is a contradiction. Thus, we proved $\|v\|_{L^{p^*}(G)} = \alpha = 1$ and $v_n \to v$ in $L^{p^*}(G)$, which shows that v is a solution of the variational problem, see (1.33).

Suppose the dichotomy case of Lemma 1.4.3 holds when applied to the sequence $d\nu_n$. We will reach a contradiction. By Remark 1.4.4 there exists a sequence of positive numbers $R_n > 0$ such that

$$\operatorname{supp} d\nu_n^1 \subset B_{R_n}(g_n), \quad \operatorname{supp} d\nu_n^2 \subset G \setminus B_{2R_n}(g_n) \qquad (1.48)$$

and

$$\lim_{n\to\infty} \left| \lambda - \int_G d\nu_n^1 \right| + \left| (1-\lambda) - \int_G d\nu_n^2 \right| = 0. \qquad (1.49)$$

Let us fix a number ϵ, such that,

$$0 < \epsilon < \lambda^{p/p^*} + (1-\lambda)^{p/p^*} - 1. \qquad (1.50)$$

Such a choice of ϵ is possible as for $0 < \lambda < 1$ and $p/p^* < 1$ we have $\lambda^{p/p^*} + (1-\lambda)^{p/p^*} - 1 > 0$. Let ϕ be a cut-off function $0 \leq \phi \in C_0^\infty(B_2(e)), \phi \equiv 1$ on $B_1(e)$. Such smooth cut-off function can be constructed explicitly in the usual manner for the gauge balls and hence by the equivalence (1.21) also for the Carnot-Carathéodory balls. Let $\phi_n = (\tau_{g_n}\phi)_{1/R_n}$, cf. (1.29), so that each ϕ_n is a smooth function with compact support, $0 \leq \phi \in C_0^\infty(B_{2R_n}(e)), \phi \equiv 1$ on $B_{R_n}(g_n)$. Then we have

$$\int_G |Xv_n|^p dH = \int_G |X(\phi_n v_n)|^p dH + \int_G |X(1-\phi_n)v_n|^p dH + \epsilon_n.$$

Note that the remainder term ϵ_n is expressed by an integral over an annuli

$$A_n = B_{2R_n}(g_n) \setminus B_{R_n}(g_n). \qquad (1.51)$$

Furthermore, we claim that

$$\epsilon_n \geq o(1) - \epsilon \int_G |Xv_n|^p dH, \text{ where } o(1) \to 0 \text{ as } n \to \infty. \qquad (1.52)$$

Indeed, using the inequality

$$(|a| + |b|)^p \leq (1+\epsilon)|a|^p + C_{\epsilon,p}|b|^p,$$

which holds for any $0 < \epsilon < 1$, $p \geq 1$ and a suitable constant $C_{\epsilon,p}$ depending on ϵ and p, we have

$$\epsilon_n = \int_{A_n} |Xv_n|^p dH - \int_{A_n} |X(\phi_n v_n)|^p dH - \int_{A_n} |X((1-\phi_n)v_n)|^p dH$$

$$\geq \int_{A_n} |Xv_n|^p (1 - \phi_n{}^p - (1-\phi_n)^p) dH - 2C_{\epsilon,p} \int_{A_n} |v_n|^p |X\phi_n|^p dH$$

$$- \epsilon \int_{A_n} |Xv_n|^p dH.$$

Since $p > 1$ and $0 \leq \psi \leq 1$ it follows $1 > \phi_n{}^p + (1-\phi_n)^p$ and thus

$$\epsilon_n \geq -C \int_{A_n} |v_n|^p |X\phi_n|^p dH - \epsilon \int_{G} |Xv_n|^p dH.$$

First we use $|X\phi_n| \leq \frac{C}{R_n}$ and then we apply Holder's inequality on A_n,

$$R_n^{-1} \|v_n\|_{L^p(A_n)} \leq R_n^{-1} |A_n|^{\frac{1}{p} - \frac{1}{p^*}} \|v_n\|_{L^{p^*}(A_n)}.$$

Since $\frac{1}{p} - \frac{1}{p^*} = \frac{1}{Q}$ and from the paragraph above (1.18)

$$|A_n| \sim R_n^Q, \tag{1.53}$$

we obtain

$$R_n^{-1} \|v_n\|_{L^p(A_n)} \leq C \|v_n\|_{L^{p^*}(A_n)}. \tag{1.54}$$

The last term in the above inequality can be estimated as follows.

$$\|v_n\|_{L^{p^*}(A_n)}^{p^*} = \int_{A_n} d\nu_n = \int_G d\nu_n - \int_{G \setminus A_n} d\nu_n$$

$$\leq \int_G d\nu_n - \int_{G \setminus A_n} d\nu_n^1 - \int_{G \setminus A_n} d\nu_n^2$$

$$= \int_G d\nu_n - \int_G d\nu_n^1 - \int_G d\nu_n^2.$$

Hence the claim (1.52) follows from

$$R_n^{-1} \|v_n\|_{L^p(A_n)} \leq C \left(\int_G d\nu_n - \int_G d\nu_n^1 - \int_G d\nu_n^2 \right)^{1/p^*} \to 0 \text{ as } n \to \infty.$$

From the definition of I and the above inequalities we have

$$\|v_n\|_{\overset{\circ}{\mathcal{D}}{}^{1,p}(G)}^p = \|\phi_n v_n\|_{\overset{\circ}{\mathcal{D}}{}^{1,p}(G)}^p + \|(1-\phi_n)v_n\|_{\overset{\circ}{\mathcal{D}}{}^{1,p}(G)}^p + \epsilon_n$$

$$\geq I \left(\|\phi_n v_n\|_{L^{p^*}(G)}^p + \|(1-\phi_n)v_n\|_{L^{p^*}(G)}^p \right) + \epsilon_n \tag{1.55}$$

$$\geq I \left(\left(\int_{B_{R_n}(g_n)} d\nu_n \right)^{p/p^*} + \left(\int_{G \setminus B_{R_n}(g_n)} d\nu_n \right)^{p/p^*} \right) + \epsilon_n \tag{1.56}$$

$$\geq I \left(\left(\int_G d\nu_n^1 \right)^{p/p^*} + \left(\int_G d\nu_n^2 \right)^{p/p^*} \right) + \epsilon_n.$$

Letting $n \to \infty$ we obtain

$$I = \lim_{n \to \infty} \|v_n\|^p_{\overset{\circ}{\mathcal{D}}{}^{1,p}(G)} \geq I\left(\lambda^{p/p^*} + (1-\lambda)^{p/p^*}\right) - \epsilon I$$

$$= I_\lambda + I_{1-\lambda} - \epsilon I,$$

which is a contradiction with the choice of ϵ in (1.50) and hence the dichotomy case of Lemma 1.4.3 cannot occur. The proof of the theorem is finished. $\qquad\qquad\square$

We end the section with the Lemmas that provide key steps in Lions' method of concentration compactness.

Lemma 1.4.3. *Suppose $d\nu_n$ is a sequence of probability measures on $\mathbf{G}$. There exists a subsequence, which we still denote by $d\nu_n$, such that exactly one of the following three cases holds.*

- *(compactness) There is a sequence $g_n \in \mathbf{G}$, $n \in \mathbb{N}$, such that for every $\epsilon > 0$ there exists $R > 0$ for which, for every n,*

$$\int_{B(g_n,R)} d\nu_n \geq 1 - \epsilon.$$

- *(vanishing) For all $R > 0$ we have*

$$\lim_{n \to \infty} \left(\sup_{g \in \mathbf{G}} \int_{B(g,R)} d\nu_n\right) = 0.$$

- *(dichotomy) There exists $\lambda, 0 < \lambda < 1$ such that for every $\epsilon > 0$ there exist $R > 0$ and a sequence (g_n) with the following property: Given $R' > R$ there exist non-negative measures $d\nu_n^1$ and $d\nu_n^2$ for which for every n we have*

$$0 \leq d\nu_n^1 + d\nu_n^2 \leq d\nu_n \qquad\qquad (1.57)$$

$$\operatorname{supp} d\nu_n^1 \subset B(g_n, R), \quad \operatorname{supp} d\nu_n^2 \subset \mathbf{G} \smallsetminus B(g_n, R') \qquad (1.58)$$

$$\left|\lambda - \int d\nu_n^1\right| + \left|(1-\lambda) - \int d\nu_n^2\right| \leq \epsilon. \qquad (1.59)$$

Proof. Let Q_n be the concentration function of $d\nu_n$. Since $\{Q_n\}$ is a sequence of non-decreasing, non-negative bounded functions on $[0, \infty)$ with $\lim_{r \to \infty} Q_n(r) = 1$ it is a locally bounded sequence in the space of functions of bounded variation on $[0, \infty)$ and thus by Helly's selection theorem there exist a subsequence, which we also denote by Q_n, and a non-decreasing function Q on $[0, \infty)$ such that

$$\lim_{n \to \infty} Q_n(r) = Q(r)$$

for r on $[0, \infty)$. Let $\lambda \overset{def}{=} \lim_{r \to \infty} Q(r)$. Clearly $0 \leq \lambda \leq 1$. If $\lambda = 0$ we have the vanishing case. Suppose $\lambda = 1$. Then there exist R_0 such that $Q(R_0) > 1/2$. For every n let g_n be such that

$$Q_n(R_0) \leq \int_{B_{R_0}(g_n)} d\nu_n + \frac{1}{n}. \tag{1.60}$$

For $0 < \epsilon < 1$ fix $R > 0$ with $Q(R) > 1 - \epsilon > \frac{1}{2}$ and take $h_n \in \boldsymbol{G}$, such that

$$Q_n(R) \leq \int_{B_R(h_n)} d\nu_n + \frac{1}{n}. \tag{1.61}$$

Then for n sufficiently large we have

$$\int_{B_R(h_n)} d\nu_n + \int_{B_{R_0}(g_n)} d\nu_n > 1 = \int_{\boldsymbol{G}} d\nu_n \tag{1.62}$$

Therefore, from the triangle inequality, $B_R(h_n) \subset B_{2\gamma R + R_0}(g_n)$ and hence

$$\int_{B_{2\gamma R + R_0}(g_n)} d\nu_n \geq \int_{B_R(h_n)} d\nu_n \geq Q_n(R) - \frac{1}{n} \to Q(R) \text{ for a.e. } R. \tag{1.63}$$

This shows that for $n \geq n_0$ and for a.e. R,

$$\int_{B_{2\gamma R + R_0}(g_n)} d\nu_n \geq 1 - \epsilon.$$

Taking a possibly larger R, we can achieve that the above holds for every n. This shows that the compactness case holds.

Finally suppose $0 < \lambda < 1$. Given $\epsilon > 0$ there exists $R > 0$ such that $Q(R) > \lambda - \epsilon$. Let be an arbitrary fixed sequence $R_n \to \infty$. We can find a subsequence of Q_n, which we also denote by Q_n, such that $\lim_{k \to \infty} Q_n(R_n) = \lambda$. This is trivial since $\lim_{n \to \infty} Q_n(R_n) = Q(R_n)$ and $\lim_{k \to \infty} Q(R_n) = \lambda$, and we can take for example Q_{n_k} sauch that $|Q(R_k) - Q_{n_k}(R_k)| < 1/k$ and $n_k > k$. Working from now on with this subsequence, there exists $n_0 = n_0(\epsilon)$ such that

$$\lambda - \epsilon < Q_n(R) < \lambda + \epsilon \qquad n \geq n_o.$$

The definition of Q_n implies we can find a sequence $(g_n) \subset G$ such that

$$\lambda - \epsilon < \int_{B_R(g_n)} d\nu_n < \lambda + \epsilon.$$

Let

$$d\nu_n^1 \overset{def}{=} d\nu_n|_{B_R(g_n)}, \quad d\nu_n^2 \overset{def}{=} d\nu_n|_{G \setminus B_{R_n}(g_n)}. \tag{1.64}$$

Obviously $0 \le d\nu_n^1 + d\nu_n^2 \le d\nu_n$ and given $R' \ge R$ for large n's we also have the condition on the supports of $d\nu_n^1$ and $d\nu_n^2$. Finally for $n \ge n_0(\epsilon)$ we have

$$\left| \lambda - \int_G d\nu_n^1 \right| + \left| (1 - \lambda) - \int_G d\nu_n^2 \right| = \left| \lambda - \int_{B_R(g_n)} d\nu_n \right|$$

$$+ \left| (-\lambda) + \int_{B_{R'}(g_n)} d\nu_n \right| \le 2\epsilon. \tag{1.65}$$

The proof of the lemma is complete.

$\square$

Remark 1.4.4. By setting $\epsilon_n \to 0$ in the above proof and taking diagonal subsequences we can also achieve

$$\operatorname{supp} d\nu_n^1 \subset B_{R_n}(g_n), \quad \operatorname{supp} d\nu_n^2 \subset G \setminus B_{2R_n}(g_n) \tag{1.66}$$

$$\lim_{n \to \infty} \left| \lambda - \int_G d\nu_n^1 \right| + \left| (1 - \lambda) - \int_G d\nu_n^2 \right| = 0. \tag{1.67}$$

Lemma 1.4.5. *Suppose* $u_n \rightharpoonup u$ *in* $\overset{o}{\mathcal{D}}{}^{1,p}(G)$, *while* $d\mu_n = |Xu_n|^p dH \rightharpoonup d\mu$ *and* $d\nu_n = |u_n|^{p^*} dH \rightharpoonup d\nu$ *weak-$*$ in measure, where* $d\mu$ *and* $d\nu$ *are bounded, non-negative measures on* G. *There exist points* $g_j \in G$ *and real numbers* $\nu_j \ge 0$, $\mu_j \ge 0$, *at most countably many different from zero, such that*

$$d\nu = |u|^{p^*} + \sum_j \nu_j \delta_{g_j},$$

$$d\mu \ge |Xu|^p dH + \sum_j \mu_j \delta_{g_j}, \quad I\nu_j^{p/p^*} \le \mu_j, \tag{1.68}$$

where I *is the constant in (1.3). In particular,*

$$\sum_j \nu_j^{p/p^*} < \infty. \tag{1.69}$$

Proof. Using the Rellich-Kondrachov type compactness, see Proposition 1.3.4 we can assume $u_n \to u$ a.e. on G. Let $v_n \overset{def}{=} u_n - u \rightharpoonup 0$ in $\overset{o}{\mathcal{D}}{}^{1,p}(G)$ and $v_n \to 0$ a.e. on G. By Corollary 1.3.6, which is a consequence of the Brézis-Lieb lemma, we have

$$d\omega_n \overset{def}{=} \left(|u_n|^{p^*} - |u|^{p^*}\right)dH = |v_n|^{p^*}dH + o(1), \tag{1.70}$$

where $o(1) \rightharpoonup 0$ in measure. Define also $d\lambda_n \overset{def}{=} |Xv_n|^p dH$. Invoking Proposition 1.3.4 we can assume, by passing to a subsequence if necessary, that $d\lambda_n \rightharpoonup d\lambda$, while $d\omega_n \rightharpoonup d\omega = d\nu - |u|^{p^*}dH$ weakly in the sense of measures for some non-negative measures $d\lambda$ and $d\omega$. For $\phi \in C_0^\infty(G)$ we have

$$
\begin{aligned}
\int_G |\phi|^{p^*}d\omega &= \lim_{n\to\infty}\int_G |\phi|^{p^*}d\omega_n = \lim_{n\to\infty}\int_G |v_n\phi|^{p^*}dH \\
&\leq I^{-p^*/p}\lim_{n\to\infty}\left(\int_G |X(v_n\phi)|^p dH\right)^{p^*/p} \\
&= I^{-p^*/p}\lim_{n\to\infty}\left(\int_G |\phi|^p |X(v_n)|^p dH\right)^{p^*/p} \\
&= I^{-p^*/p}\left(\int_G |\phi|^p d\lambda\right)^{p^*/p},
\end{aligned}
$$

using that $\int_G |v_n|^p |X(\phi)|^p dH \to 0$, which follows either from Hölder's inequality and $v_n \rightharpoonup 0$ in $L^{p^*}(G)$,

$$\left(\fint_B |v_n|^p |X(\phi)|^p dH\right)^{1/p} \leq \left(\fint_B |v_n|^{p^*} |X(\phi)|^p dH\right)^{1/p^*}, \qquad \phi \in C_0^\infty(B),$$

or from Rellich-Kondrachov compact embedding Theorem 1.3.1. Thus, we come to the following reverse Holder inequality

$$\left(\int_G |\phi|^{p^*} d\omega\right)^{1/p^*} \leq I^{-1/p}\left(\int_G |\phi|^p d\lambda\right)^{1/p}.$$

Now, by Lemma 1.4.6 we obtain

$$d\nu = |u|^{p^*}dH + \sum_{j\in J}\nu_j\delta_{g_j} \text{ and } d\lambda \geq I\sum_{j\in J}\nu_j^{p/p^*}.$$

From the weak convergence $u_n \rightharpoonup u$ in $\overset{o}{\mathcal{D}}{}^{1,p}(G)$ it follows $d\mu_n - d\lambda_n = |Xu|^p dH + o(1)$ which combined with the inequality above gives the desired estimate for $d\mu$. $\qquad\square$

We end the section with a technical lemma involving a reverse Hölder inequality, which was used at the end of the above proof of Lemma 1.4.5.

Lemma 1.4.6. *Let $d\lambda$ and $d\omega$ be two non-negative measures on $\boldsymbol{G}$ satisfying for some C_0 and $1 \leq p < q \leq \infty$*

$$\left(\int_G |\phi|^q d\omega\right)^{1/q} \leq C_0 \left(\int_G |\phi|^p d\lambda\right)^{1/p}, \text{ for every } \phi \in C_0^\infty(\boldsymbol{G}). \quad (1.71)$$

Then there exist an at most countable set J s.t.

$$d\omega = \sum_{j \in J} \nu_j \delta_{g_j}, \qquad d\lambda \geq C_0^{-p} \sum_{j \in J} \nu_j^{p/q} \delta_{g_j}$$

for some numbers $\nu_j \geq 0$ and $g_j \in \boldsymbol{G}$, $j \in J$. In particular $\sum_{j \in J} \nu_j^{p/q} < \infty$.

Proof. Clearly $d\omega$ is absolutely continuous w.r.t. $d\lambda$ and thus $d\omega = f d\lambda$, where $f \in L^1(d\lambda), f \geq 0$. Furthermore f is bounded since $\omega(E) \leq C_0 \lambda(E)^{q/p}$ and $q/p > 1$. Decomposing $d\lambda = g d\omega + \sigma$ with $\sigma \perp \omega$, it is enough to prove (1.71) for $d\lambda \overset{def}{=} g d\omega$ since $d\sigma$ is a non-negative measure. Let $d\omega_k \overset{def}{=} g^\alpha 1_{\{g \leq k\}} d\omega$ with $\alpha \overset{def}{=} \frac{q}{q-p}$. We are going to prove that $d\omega_k$ is given by a finite number of Dirac measures. This will prove that $1_{\{g \leq k\}} d\omega = g^{-\alpha} d\omega_k$ is a sum of finite number of Dirac measures for all $k < \infty$ and letting $k \to \infty$ the claim on $d\omega$ will be proved since $\omega(\{g = \infty\}) = 0$ as $g \in L^1(d\omega)$ Take $\phi = g^{\frac{1}{q-p}} 1_{\{g \leq k\}} \psi$, ψ an arbitrary bounded measurable function. Then we have

$$\int_G |\phi|^q d\omega = \int_G g^{\frac{q}{q-p}} 1_{\{g \leq k\}} |\psi|^q d\omega = \int_G |\psi|^q d\omega_k.$$

On the other hand

$$\int_G |\phi|^q d\lambda = \int_G g^{\frac{p}{q-p}} 1_{\{g \leq k\}} |\psi|^p d\omega = \int_G g^{\frac{q}{q-p}} 1_{\{g \leq k\}} |\phi|^p d\omega$$

$$= \int_G |\psi|^p d\omega_k.$$

In other words we showed that (1.71) holds also for $d\omega_k$ on both sides of the inequality. Thus for any Borel set A we have

$$d\omega_k(A)^{1/q} \leq C_0 d\omega_k(A)^{1/p}$$

and therefore either $d\omega_k(A) = 0$ or $d\omega_k(A) \geq C_0^{-\frac{p}{q-p}} > 0$. In particular for any $g \in \boldsymbol{G}$ we have either $d\omega_k(\{g\}) \geq C_0^{-\frac{p}{q-p}}$ or there exists an $\epsilon > 0$ such that $d\omega_k(B_\epsilon(g)) = 0$ since $d\omega_k(\{g\}) = \lim_{\epsilon \to \infty} d\omega_k(B_\epsilon(g))$. From the boundedness of the measures we conclude that $d\omega_k$ is a sum of at most finite number of Dirac measures. The proof is finished since the second part follows trivially from the first having in mind the reverse Holder inequality. $\square$

1.5   The best constant in the presence of symmetries

We consider here the variational problem (1.3), but we restrict the class of test functions.

Definition 1.5.1. Let G be a Carnot group with Lie algebra $\mathfrak{g} = V_1 \oplus V_2 \cdots \oplus V_n$. We say that a function $U : G \to \mathbb{R}$ has *partial symmetry* with respect to g_o if there exist an element $g_o \in G$ such that for every $g = \exp(\xi_1 + \xi_2 + \cdots + \xi_n) \in G$ one has

$$U(g_o g) = u(|\xi_1(g)|, |\xi_2(g)|, \ldots, \xi_n(g)),$$

for some function $u : [0, \infty) \times [0, \infty) \ldots [0, \infty) \times V_n \to \mathbb{R}$.

A function U is said to have *cylindrical symmetry* if there exist $g_o \in G$ and $\phi : [0, \infty) \times [0, \infty) \cdots \times [0, \infty) \to \mathbb{R}$ for which

$$U(g_o g) = \phi(|\xi_1(g)|, |\xi_2(g)|, \ldots, |\xi_n(g)|),$$

for every $g \in G$.

We define also the spaces of partially symmetric and cylindrically symmetric function $\overset{o}{\mathcal{D}}_{ps}{}^{1,p}(G)$ and $\overset{o}{\mathcal{D}}_{cyl}{}^{1,p}(G)$, respectively, as follows

$$\overset{o}{\mathcal{D}}_{ps}{}^{1,p}(G) \overset{def}{=} \{u \in \overset{o}{\mathcal{D}}{}^{1,p}(G) : u(g) = u(|\xi_1(g)|, \ldots, |\xi_{n-1}(g)|, \xi_n(g))\}, \tag{1.72}$$

and

$$\overset{o}{\mathcal{D}}_{cyl}{}^{1,p}(G) \overset{def}{=} \{u \in \overset{o}{\mathcal{D}}{}^{1,p}(G) : u(g) = u(|\xi_1(g)|, \ldots, |\xi_n(g)|)\}. \tag{1.73}$$

The effect of the symmetries, see also [126], is manifested in the fact that if the limit measure given by Lemma 1.4.5 concentrates at a point, then it must concentrates on the whole orbit of the group of symmetries. Therefore, in the cylindrical case there could be no points of concentration except at the origin, while in the partially-symmetric case the points of concentration lie in the center of the group.

Theorem 1.5.2.

a) *The norm of the embedding $\overset{o}{\mathcal{D}}_{ps}{}^{1,p}(G) \subset L^{p^*}(G)$ is achieved.*

b) *The norm of the embedding $\overset{o}{\mathcal{D}}_{cyl}{}^{1,p}(G) \subset L^{p^*}(G)$ is achieved.*

In order to rule out the dichotomy case in the first part of the theorem we prove the following lemma.

Lemma 1.5.3. *Under the conditions of Lemma 1.4.3, the points g_n in the dichotomy part can be taken from the center of the group.*

Proof. Define the concentration function of ν_n by

$$Q_n^{ps} \stackrel{def}{=} \sup_{h \in C(G)} \left(\int_{B_r(h)} d\nu_n \right). \tag{1.74}$$

The rest of the proof is identical to the proof of Lemma 1.4.3, with the remark that in the dichotomy part, the definition of Q_n^{ps} shows that the points g_n can be taken to belong to the center. $\qquad\square$

We turn to proof of Theorem 1.5.2.

Proof. We argue as in Theorem 1.4.2.

a) Finding the norm of the embedding $\overset{o}{\mathcal{D}}_{ps}{}^{1,p}(G) \subset \overset{o}{\mathcal{D}}{}^{1,p}(G)$ leads to the following variational problem,

$$I^{ps} \equiv I_1^{ps} \stackrel{def}{=} \inf \left(\int_G |Xu|^p : u \in \overset{o}{\mathcal{D}}_{ps}{}^{1,p}(G), \int_G |u|^{p^*} = 1 \right). \tag{1.75}$$

Let us take a minimizing sequence (u_n), i.e.,

$$\int_G |u_n|^{p^*} = 1 \text{ and } \int_G |Xu_n|^p \underset{n \to \infty}{\to} I^{ps}. \tag{1.76}$$

It is clear that $\overset{o}{\mathcal{D}}_{ps}{}^{1,p}(G)$ is invariant under the dilations (1.29). Using the Baker-Campbell-Hausdorff formula it is easy to see that $\overset{o}{\mathcal{D}}_{ps}{}^{1,p}(G)$ is invariant, also, under the translations (1.28) by elements in the center, $C(G)$, of G. In order to extract a suitable dilated and translated subsequence of $\{u_n\}$ we have to make sure that we translate always by elements belonging to $C(G)$. In order to achieve this we define the concentration function of u_n as

$$\hat{Q}_n^{ps}(r) \stackrel{def}{=} \sup_{h \in C(G)} \left(\int_{B_r(h)} |u_n|^{p^*} dH \right). \tag{1.77}$$

We can fix $r_n > 0$ and $h_n \in C(G)$, such that (1.36), and (1.37) hold. Define the sequence $\{v_n\}$ as in (1.38), hence $v_n \in \overset{o}{\mathcal{D}}_{ps}{}^{1,p}(G)$. Using the translations and scaling as before we obtain (1.39). At this point we can apply Lemma 1.5.3. The case of vanishing is ruled out from the normalization (1.39) of the sequence $\{v_n\}$. Suppose we have dichotomy. Let us take a sequence $R_n > 0$ such that (1.48) and (1.49) hold. We choose a cut-off function, ϕ, from the space $\overset{o}{\mathcal{D}}_{ps}{}^{1,p}(G)$, satisfying also

$$\text{supp } \phi \subset \Omega_2(e), \qquad \text{and} \qquad \phi \equiv 1 \quad \text{on } \Omega_1(e), \tag{1.78}$$

where $\Omega_r(g)$ denotes a gauge ball centered at g and radius r, i.e.,

$$\Omega_r(g) = \{h \in G : N(h^{-1}g) < r\}. \tag{1.79}$$

This can be done by setting $\phi = \eta(N(g))$, where $\eta(t)$ is a smooth function on the real line, supported in $|t| < 2$ and $\eta \equiv 1$ on $|t| \leq 1$. We define the cut off functions ϕ_n as before, $\phi_n = (\tau_{g_n}\phi)_{1/R_n}$. From the Baker-Campbell-Hausdorff they have partial symmetry with respect to the identity since $g_n \in C(G)$, see Lemma 1.5.3, and the gauge is a function with partial symmetry G. By letting

$$A_n = \Omega_{2R_n}(g_n) \smallsetminus \Omega_{R_n}(g_n) \tag{1.80}$$

and noting that

$$|A_n| \sim R_n^Q, \tag{1.81}$$

we see that (1.52) holds. Now, from the definition of I_{ps}, having in mind that ϕ_n and v_n have partial symmetry with respect to the identity, we obtain

$$
\begin{aligned}
\|v_n\|_{\overset{o}{\mathcal{D}}^{1,p}(G)}^p &= \|\phi_m v_n\|_{\overset{o}{\mathcal{D}}^{1,p}(G)}^p + \|(1-\phi_n)v_n\|_{\overset{o}{\mathcal{D}}^{1,p}(G)}^p + \epsilon_n \\
&\geq I_{ps}\left(\|\phi_m v_n\|_{L^{p^*}(G)}^p + \|(1-\phi_n)v_n\|_{L^{p^*}(G)}^p\right) + \epsilon_n \\
&\geq I_{ps}\left(\left(\int_{B_{R_n}(g_n)} d\nu_n\right)^{p/p^*} + \left(\int_{G \smallsetminus B_{R_n}(g_n)} d\nu_n\right)^{p/p^*}\right) + \epsilon_n \\
&\geq I_{ps}\left(\left(\int_G d\nu_n^1\right)^{p/p^*} + \left(\int_G d\nu_n^2\right)^{p/p^*}\right) + \epsilon_n \\
&\geq I_{ps}\left(\lambda^{p/p^*} + (1-\lambda)^{p/p^*}\right) + \epsilon_n.
\end{aligned}
$$

Letting $n \to \infty$ we come to

$$\lim_{n\to\infty} \|v_n\|_{\overset{o}{\mathcal{D}}^{1,p}(G)}^p \geq I_{ps}\left(\lambda^{p/p^*} + (1-\lambda)^{p/p^*}\right) > I_{ps}, \tag{1.82}$$

since $0 < \lambda < 1$ and $\frac{p}{p^*} < 1$ This contradicts $\|v_n\|_{\overset{o}{\mathcal{D}}^{1,p}(G)}^p \to I^{ps}$ as $n \to \infty$, which shows that the dichotomy case of Lemma 1.5.3 cannot occur. Hence the compactness case holds. As in Theorem 1.4.2 we see that

$$\int_G d\nu = 1.$$

Next, we apply Lemma 1.4.5, with I replaced by I_{ps}. The important fact here is that the partial symmetry of the sequence $\{v_n\}$ implies the points of concentration of $d\nu$ must be in the center of the group if they occur.

Having this in mind and also the definitions of the concnetration functions, we can justify the validity of (1.47), and finish the proof of part a).

b) The vanishing case is ruled out by using the dilation (but not translation because of the symmetries) invariance and normalizing the minimizing sequence with the condition $Q_n(1) = 1/2$, see (1.39).

Suppose that the dichotomy case occurs. We shall see that this leads to a contradiction. The points $\{g_n\}$ in the dichotomy part of Lemma 1.4.3 must be a bounded sequence. If not, let $\epsilon = \lambda/2$ and R be the as in the Lemma. Due to the invariance by rotations in the layers of the functions v_n and the Haar measure dH, which is just the Lebesgue measure, for any arbitrarily fixed natural number, N_o, we can find a point g_n and N_o points on the orbit of g_n under rotations in one of the layers, such that the balls with radius R centered at all these points do not intersect. This leads to a contradiction since the integral of the probability measure $d\nu_n$ over each of these balls is greater than $\lambda/2$. Thus, $\{g_n\}$ is a bounded sequence. This is however impossible since $d\nu_n$ are probability measures.

Therefore, the compactness case holds. Exactly as in the dichotomy part we see that the sequence $\{g_n\}$ is a bounded sequence. This amounts to saying, using the triangle inequality, that we can take all of them at the identity. We conclude that

$$\int_G d\nu = 1.$$

We can finish the proof as in Theorem 1.4.2. □

1.6 Global regularity of weak solutions

1.6.1 *Global boundedness of weak solutions*

Let $1 < p < Q$ and denote by p^* the Sobolev conjugate $p^* = \frac{pQ}{Q-p}$ and by p' the Hölder conjugate $p' = \frac{p}{p-1}$. Let $u \in \overset{o}{\mathcal{D}}{}^{1,p}(\Omega)$ be a weak solution in an open set, not necessarily bounded, $\Omega \subset G$ of the equation (1.5). Weak solution means that for every $\phi \in C_o^\infty(\Omega)$ we have

$$\int_\Omega |Xu|^{p-2} < Xu, X\phi > dH = \int_\Omega |u|^{p^*-2} u \, \phi \, dH. \qquad (1.83)$$

Note that $u^{p^*-1} \in L^{\frac{p^*}{p^*-1}}(\Omega) = L^{(p^*)'}$. From the definition of $\overset{o}{\mathcal{D}}{}^{1,p}(\Omega)$ we obtain that (1.83) holds for every $\phi \in \overset{o}{\mathcal{D}}{}^{1,p}(\Omega)$. The main result of this section is that weak solutions as above are bounded functions. In the following

Theorem we prove a more general result. For the ordinary Laplacian in a bounded domain, Brezis and Kato [31] established a result similar to part a) of Theorem 1.6.1, see [33] for b). The proofs rely on a suitable modification of the test function and truncation ideas introduced in [152] and [131]. We note that in all these results the solution is assumed a-priori to be in the space $L^p(\Omega)$ since this is part of the definition of the considered Sobolev spaces. This is not true in the Sobolev spaces $\overset{o}{\mathcal{D}}{}^{1,p}(\Omega)$ we consider, since we include only the L^p norm of the horizontal gradient in our definition. Therefore, the results here are not exactly the same, besides that we are working on a Carnot group. Subsequently Serrin's ideas were generalized to the subelliptic setting in [38], and also in different forms in [91], [92], [119], [173].

Theorem 1.6.1. *Let* $u \in \overset{o}{\mathcal{D}}{}^{1,p}(\Omega)$ *be a weak solution to the equation*

$$\sum_{i=1}^{m} X_i(|Xu|^{p-2}X_i u) = -V\,|u|^{p-2}u \qquad in \quad \Omega, \qquad (1.84)$$

i.e., for every $\phi \in C_o^\infty(\Omega)$ *we have*

$$\int_\Omega |Xu|^{\,p-2} < Xu, X\phi > dH = \int_\Omega V|u|^{p-2}u\,\phi\,dH. \qquad (1.85)$$

a) If $V \in L^{Q/p}(\Omega)$, *then* $u \in L^q(\Omega)$ *for every* $p^* \le q < \infty$.
b) If $V \in L^t(\Omega) \cap L^{Q/p}(\Omega)$ *for some* $t > \frac{Q}{p}$, *then* $u \in L^\infty(\Omega)$.

Proof. The assumption $V \in L^{Q/p}(\Omega)$ together with the Folland-Stein inequality shows that (1.85) holds true for any $\phi \in \overset{o}{\mathcal{D}}{}^{1,p}(\Omega)$. This follows from the density of the space $C_o^\infty(\Omega)$ in the space $\overset{o}{\mathcal{D}}{}^{1,p}(\Omega)$, which will allow to put the limit in $\overset{o}{\mathcal{D}}{}^{1,p}(\Omega)$ of a sequence $\phi_n \in C_o^\infty(\Omega)$ in the left-hand side of (1.85). On the other hand, the Folland-Stein inequality implies that $\phi_n \to \phi$ in $L^{p^*}(\Omega)$. Set $t_o = \dfrac{Q}{p}$ and its Hölder conjugate $t_o' = \dfrac{t_o}{t_o - 1}$. An easy computation gives

$$\frac{1}{t_o} + \frac{p-1}{p^*} = 1 - \frac{1}{p^*} = \frac{1}{(p^*)'}.$$

Hölder's inequality shows that $V|u|^{p-2}u \in L^{(p^*)'}(\Omega)$, which allows to pass to the limit in the right-hand side of (1.85). We turn to the proofs of a) and b).

a) It is enough to prove that $u \in \overset{o}{\mathcal{D}}{}^{1,p}(\Omega) \cap L^q(\Omega)$, $q \geq p^*$ implies $u \in L^{\kappa q}$ with $\kappa = \dfrac{p^*}{p} > 1$. Let $G(t)$ be a Lipschitz function on the real line, and set

$$F(u) = \int_0^u |G'(t)|^p \, dt. \tag{1.86}$$

Clearly, F is a differentiable function with bounded and continuous derivative. From the chain rule, see for ex. [74], $F(u) \in \overset{o}{\mathcal{D}}{}^{1,p}(\Omega)$ is a legitimate test function in (1.85). The left-hand side, taking into account $F'(u) = |G'(u)|^p$, can be rewritten as

$$\int_\Omega |Xu|^{p-2} < Xu, XF(u) > dH = \int_\Omega |XG(u)|^p.$$

The Folland-Stein inequality (1.1) gives

$$\int_\Omega |Xu|^{p-2} < Xu, XF(u) > dH \geq S_p \left(\int_\Omega |G(u)|^{p^*} \right)^{p/p^*}. \tag{1.87}$$

Let us choose $G(t)$ in the following way,

$$G(t) = \begin{cases} sign\,(t)\,|t|^{\frac{q}{p}} & \text{if } 0 \leq |t| \leq l, \\ l^{\frac{q}{p}-1} t & \text{if } l < |t|. \end{cases}$$

From the power growth of G, besides the above properties, this function satisfies also

$$|u|^{p-1}|F(u)| \leq C(q)|G(u)|^p \leq C(q)|u|^q. \tag{1.88}$$

The constant $C(q)$ depends also on p, but this is a fixed quantity for us. At this moment the value of $C(q)$ is not important, but an easy calculation shows that $C(q) \leq C\,q^{p-1}$ with C depending on p. We will use this in part b). Note that $pt'_o = p^*$. Let $M > 0$ to be fixed in a moment and estimate the integral in the right-hand side of (1.85) as follows.

$$\int_\Omega V|u|^{p-2}uF(u)\,dH$$

$$= \int_{(|V|\leq M)} V|u|^{p-2}uF(u)\,dH + \int_{(|V|>M)} V|u|^{p-2}uF(u)\,dH$$

$$\leq M\int_{(|V|\leq M)} |u|^{p-1}F(u)\,dH + \left(\int_{(|V|>M)} |V|^{t_o}dH\right)^{\frac{1}{t_o}}\left(\int_\Omega (|u|^{p-1}F(u))^{t'_o}dH\right)^{\frac{1}{t'_o}}$$

$$\leq C(q)M\int_\Omega |G(u)|^p\,dH + \left(\int_{(|V|>M)} |V|^{t_o}\right)^{\frac{1}{t_o}}\left(\int_\Omega |G(u)|^{p^*}\,dH\right)^{\frac{p}{p^*}}.$$

$$(1.89)$$

At this point we fix an M sufficiently large, so that

$$C(q)\left(\int_{(|V|>M)} |V|^{t_o}\,dH\right)^{\frac{1}{t_o}} \leq \frac{S_p}{2},$$

which can be done because $V \in L^{t_o}$. Putting together (1.87) and (1.89) we come to our main inequality

$$\frac{S_p}{2}\left(\int_\Omega |G(u)|^{p^*}\,dH\right)^{\frac{p}{p^*}} \leq C(q)M\int_\Omega |G(u)|^p\,dH \leq C(q)M\int_\Omega |u|^q\,dH.$$

By the Fatou and Lebesgue dominated convergence theorems we can let l in the definition of G to infinity and obtain

$$\frac{S_p}{2}\left(\int_\Omega |u|^{\frac{p^*}{p}q}\,dH\right)^{\frac{p}{p^*}} \leq C(q)M\int_\Omega |u|^q\,dH.$$

The proof of a) is finished.

b) It is enough to prove that the $L^q(\Omega)$ norms of u are uniformly bounded by some sufficiently large but fixed L^{q_o} norm of u, $q_o \geq p^*$, which is finite from a). We shall do this by iterations [131]. We use the function $F(u)$ from part a) in the weak form (1.85) of our equation. The left-hand side is estimated from below as before, see (1.87). This time, though, we use Hölder's inequality to estimate from above the right-hand side,

$$\int_\Omega V|u|^{p-2}uF(u)\,dH \leq \|V\|_t \|\,|u|^{p-1}F(u)\|_{t'}$$

$$\leq \|V\|_t \|C(q)|G(u)|^p\|_{t'} \leq C(q)\|V\|_t \|u\|_{qt'}^q. \quad (1.90)$$

With the estimate from below we come to
$$S_p \|G(u)\|_{p^*}^p \leq C(q) \|V\|_t \|u\|_{qt'}^q.$$
Letting $l \to \infty$ we obtain
$$\left\| |u|^{q/p} \right\|_{p^*}^p \leq \frac{C(q)}{S_p} \|V\|_t \|u\|_{qt'}^q. \tag{1.91}$$

Let $\delta = \dfrac{p^*}{p\,t'}$. The assumption $t > \dfrac{Q}{p}$ implies $\delta > 1$, since the latter is

equivalent to $t' < \dfrac{p^*}{p} = t'_o$ from $t_o = \dfrac{Q}{p}$. With this notation we can

rewrite (1.91) as
$$\|u\|_{\delta q t'} \leq \left[\frac{C(q)}{S_p} \right]^{\frac{1}{q}} \|V\|_t^{\frac{1}{q}} \|u\|_{qt'}. \tag{1.92}$$

Recall that $C(q) \leq C q^{p-1}$. At this point we define $q_o = p^* t'$ and $q_k = \delta^k q_o$, and after a simple induction we obtain
$$\|u\|_{q_k} \leq \prod_{j=0}^{k-1} \left[C q_j^{p-1} \right]^{\frac{1}{q_j}} \|V\|_t^{\sum_{j=0}^{k-1} \frac{1}{q_j}} \|u\|_{q_o}. \tag{1.93}$$

Let us observe that the right-hand side is finite,
$$\sum_{j=0}^{\infty} \frac{1}{q_j} = \frac{1}{q_o} \sum_{j=1}^{\infty} \frac{1}{\delta^j} < \infty \qquad \text{and} \qquad \sum_{j=1}^{\infty} \frac{\log q_j}{q_j} < \infty, \tag{1.94}$$
because $\delta > 1$. Letting $j \to \infty$ we obtain
$$\|u\|_{\infty} \leq C \|u\|_{q_o}.$$
$\square$

Remark 1.6.2. When Ω is a bounded open set we have trivially $V \in L^{Q/p}(\Omega)$ when $V \in L^t(\Omega)$, $t > \dfrac{Q}{p}$. Also, in this case one can obtain a uniform estimate of the $L^{\infty}(\Omega)$ norm of u by its $L^{p^*}(\Omega)$ norm, which does not depend on the distribution function of V, as we have in the above Theorem. This can be achieved even in the unbounded case, assuming only $V \in L^{Q/p}(\Omega)$ if we required though $u \in L^p(\Omega)$.

With the above Theorem we turn to our original equation (1.5).

Theorem 1.6.3. *Let $1 < p < Q$ and $\Omega \subset \mathbf{G}$ be an open set. If $u \in \overset{o}{\mathcal{D}}{}^{1,p}(\Omega)$ is a weak solution to equation*
$$\mathcal{L}_p u = \sum_{j=1}^{n} X_j(|Xu|^{p-2} X_j u) = -|u|^{p^*-2} u \qquad \text{in} \quad \Omega,$$
then $u \in L^{\infty}(\Omega)$.

Proof. We define $V = |u|^{p^*-p}$. From the Folland-Stein inequality we have $u \in L^{p^*}(\Omega)$ and thus $V \in L^{\frac{p^*}{p^*-p}}(\Omega)$. Since $\dfrac{p^*}{p^*-p} = \dfrac{Q}{p}$, part (a) of Theorem 1.6.1 shows that $u \in L^q(\Omega)$ for $p^* \leq q < \infty$. Therefore $V \in L^{\frac{q}{p^*-p}}(\Omega)$ for any such q and thus by part (b) of the same Theorem we conclude $u \in L^{\infty}(\Omega)$. □

1.6.2   The Yamabe equation - $\mathcal{C}^{\infty}$ regularity of weak solutions

In this section we explain how one obtains the $\mathcal{C}^{\infty}$ smoothness of the extremals of the L^2 Folland-Stein embedding, i.e., we prove the $\mathcal{C}^{\infty}$ smoothness of the non-negative weak solutions of the Yamabe equation.

Thus, given a Carnot group G we consider the *sub-Laplacian* associated with the system $\{X_1, \ldots, X_m\}$, which is the second-order partial differential operator given by

$$\mathcal{L} = -\sum_{j=1}^{m} X_j^* X_j = \sum_{j=1}^{m} X_j^2,$$

see (1.16). By the assumption on the Lie algebra the system $\{X_1, \ldots, X_m\}$ satisfies the well-known finite rank condition, therefore thanks to Hörmander's theorem [93] the operator $\mathcal{L}$ is hypoelliptic, see also [144]. However, it fails to be elliptic, and the loss of regularity is measured in a precise way, which becomes apparent in the corresponding sub-elliptic regularity estimates, by the number of layers r in the stratification of $\mathfrak{g}$.

On a stratified group G we have the analogues of Sobolev and Lipschitz spaces, the so called non-isotropic Sobolev and Lipschitz spaces [67], [65], defined as follows. Using $\mathcal{BC}$ for the space of all bounded continuous functions and $|h|$ for the homogeneous norm of $h \in G$ we let

$$S^{k,p}(G) = \{f \in L^p(G, dH) : X_I f \in L^p(G, dH) \text{ for } |I| \leq k\},$$
$$k \in \mathbb{N} \cup \{0\},$$

$$\Gamma_\alpha(G) = \{f \in \mathcal{BC} : \sup_{g,h \in G, h \neq e} \frac{|f(gh) - f(g)|}{|h|^\alpha} < \infty\}, \quad 0 < \alpha < 1,$$

$$\Gamma_1(G) = \{f \in \mathcal{BC} : \sup_{g,h \in G, h \neq e} \frac{|f(gh) + f(gh^{-1}) - 2f(g)|}{|h|} < \infty\},$$

$$\Gamma_{k+\alpha}(G) = \{f \in \mathcal{BC} : X_I f \in \Gamma_\alpha(G) \text{ for every } |I| \leq k\},$$
$$0 < \alpha \leq 1, \ k \in \mathbb{N}.$$

$$(1.95)$$

As usual, in the above definition we used the notation X_I to denote $X_{i_1} X_{i_2} \ldots X_{i_l}$, $1 \le i_j \le m$, where I is the multi-index $I = (i_1, \ldots, i_l)$ with $|I| = l$. In other words, $X_I f$ is obtained by differentiating f along the first layer $|I|$ times. We equip the above spaces with norms turning them into Banach spaces, for example, for $0 < \alpha < 1$,

$$\|f\|_{\Gamma_\alpha(G)} = \|f\|_{L^\infty(G)} + \sup_{g,h \in G} \frac{|f(gh) - f(g)|}{|h|^\alpha}.$$

The local versions of the above spaces are defined in the obvious way. It is possible to define non-isotropic Sobolev spaces for non-integer k's using fractional powers of the sub-Laplacian [65], but we will not do this.

The non-isotropic spaces are known to satisfy the following inclusion properties when related to the corresponding isotropic spaces in Euclidean space.

Theorem 1.6.4. *[67] Let G be a Carnot group of step r and $W_{loc}^{k,p}(G)$, and $\Lambda_{\alpha,loc}(G)$ denote correspondingly the isotropic (the "usual") Sobolev and Lipschitz spaces on the Euclidean space $\mathfrak{g}$. The following embeddings hold true.*

$$W_{loc}^{k,p}(G) \subset S_{loc}^{k,p}(G) \subset W_{loc}^{k/r,p}(G) \text{ for } 1 < p < \infty, \ k \in \mathbb{N}$$
$$\Lambda_{\alpha,loc}(G) \subset \Gamma_\alpha(G) \subset \Lambda_{\alpha/r,loc}(G), \qquad \alpha > 0. \tag{1.96}$$

Sometimes it is useful to have in mind the following characterization of the non-isotropic Lipschitz spaces on Carnot groups.

Theorem 1.6.5. *[118] $f \in \Gamma_\alpha(G)$ if and only if $f(g_t) \in \Lambda_\alpha$ for every horizontal curve g_t starting from any $g_0 \in G$.*

A curve is called horizontal if $\frac{d}{dt} g_t \in \text{span}\{X_1, \ldots, X_m\}$.

For completeness, we also state the generalization of the Sobolev embedding theorem relating the non-isotropic spaces.

Theorem 1.6.6. *[67]*

a) $S_{loc}^{k,p}(G) \subset S_{loc}^{l,q}(G)$ *for any* $1 < p < q < \infty$ *and* $l = k - Q(1/p - 1/q) \ge 0$.

b) $S_{loc}^{k,p}(G) \subset \Gamma_\alpha(G)$ *when* $\alpha = k - (Q/p) > 0$.

The main regularity result concerning the sub-Laplacian acting on non-isotropic Sobolev or Lipschitz spaces is the following Theorem.

Theorem 1.6.7. *[67; 65] Let G be a Carnot group of step r. If $\mathcal{L}u$ is in one of the spaces $S_{loc}^{k,p}(G)$ or $\Lambda_{\alpha,loc}(G)$ for some $\alpha > 0$, $1 < p < \infty$ and $k \in \mathbb{N} \cup \{0\}$, then u is in $S_{loc}^{k+2,p}(G)$ or $\Lambda_{\alpha+(2/r),loc}(G)$, respectively.*

Remark 1.6.8. The above theorem is valid for any system of vector fields that satisfy the Hörmander condition of order r, see [66] for a general overview and further details.

The Hölder regularity of weak solutions of equation (1.97) follows from a suitable adaptation of the classical De Giorgi-Nash-Moser result. The higher regularity when $p = 2$ follows by an iteration argument based on Theorem 1.6.7.

Theorem 1.6.9. *Let Ω be an open set in a Carnot group $\mathbf{G}$.*

a) *If $u \in \overset{o}{\mathcal{D}}{}^{1,p}(\Omega)$ is a non-negative weak solution to the equation*

$$\sum_{i=1}^{m} X_i(|Xu|^{p-2}X_i u) = -V\, u^{p-1} \qquad in \quad \Omega \qquad (1.97)$$

and $V \in L^t(\Omega)$ for some $t > \frac{Q}{p}$, then u satisfies the Harnack inequality: for any Carnot-Carathédory (or gauge) ball $B_{R_0}(g_0) \subset \Omega$ there exists a constant $C_0 > 0$ such that

$$\operatorname*{esssup}_{B_R} u \le C_0 \operatorname*{essinf}_{B_R} u, \qquad (1.98)$$

for any Carnot-Carathédory (or gauge) ball $B_R(g)$ such that $B_{4R}(g) \subset B_{R_0}(g_0)$.

b) *If $u \in \overset{o}{\mathcal{D}}{}^{1,p}(\Omega)$ is a weak solution to (1.97) and $V \in L^t(\Omega) \cap L^{Q/p}(\Omega)$, then $u \in \Gamma_\alpha(\Omega)$ for some $0 < \alpha < 1$.*

c) *If $u \in \overset{o}{\mathcal{D}}{}^{1,2}(\Omega)$ is a non-negative weak solution, which does not vanish identically, of the Yamabe equation on the domain Ω,*

$$\mathcal{L}u = -u^{2^*-1}, \qquad (1.99)$$

then $u > 0$ and $u \in \mathcal{C}^\infty(\Omega)$.

Part a) is a particular case of the Harnack inequality of [[38], Theorem 3.1], where the result is proven for more general Hörmander type vector fields and a non-linear sub-elliptic equation modeled on (1.97). Notice that [38] uses the Sobolev spaces $\mathcal{S}^{1,p}(\Omega)$ rather than $\overset{o}{\mathcal{D}}{}^{1,p}(\Omega)$. In particular the Harnack inequality is proven for solutions which are locally in $\mathcal{S}^{1,p}(\Omega)$. The first step in the proof of the Harnack inequality is the establishment of the local boundedness of weak solutions (without the non-negativity assumption). Under the assumption $V \in L^t(\Omega) \cap L^{Q/p}(\Omega)$, $t > \frac{Q}{p}$, we have even global boundedness $u \in L^\infty(\Omega)$, a detailed proof of which was given in Theorem 1.6.1.

The claim of part b) is contained in [[38], Theorem 3.35] as a special case.

For the proof of part c) we observe first that by part b) u is Hölder continuous, while part a) gives that $u > 0$. Furthermore, by Theorem 1.6.3 there exists $M > 0$ such that $u \leq M < \infty$. Thus, if we define $f(u) = u^{2^*-1}$, the fact that $0 < u \leq M$ allows us to use that $f \in C^\infty((0, +\infty))$. A bootstrap argument using the Lipschitz case of Theorem 1.6.7 proves the claim in part c).

In the case of a group of Heisenberg type, cf. Definition 2.2.1, parts b) and c) can be proved also with the help of the Hopf boundary Lemma 3.2.2 in a manner independent of the Harnack inequality. Since this is the most important case in this book, we present the argument. By Theorem 1.6.1 u is a bounded function, thus $u \in S_{loc}^{0,p}(G) = L_{loc}^p(G)$ for any $p > 0$. Now, Theorem 1.6.7 shows $u \in S_{loc}^{2,p}(G)$ for any $p > 0$. By the embedding Theorem 1.6.6 it follows u is a Lipschit continuous function in the sense of non-isotropic Lipschit spaces, $u \in \Gamma_\alpha(G)$ when $\alpha = 2 - (Q/p) > 0$. In particular, u is a continuous function by Theorem 1.6.4 since the Heisenberg type groups are Carnot groups. Thus, the set $\{u > 0\}$ is an open subset of G. Iterating this argument and using Theorem 1.6.4 we see that $u \in C^\infty$ smooth function on the set where it is positive, while being of class $\Gamma_{loc}^{2,\alpha}(G)$ on the whole group using that the Heisenberg type groups are two-step Carnot groups. Again, by Theorem 1.6.4 we conclude that u is a continuously differentiable function. Applying the Hopf Lemma 3.2.2 on the set where u is positive shows that u cannot vanish, i.e., $u > 0$ and $u \in C^\infty(\Omega)$. The existence of an interior gauge ball needed to apply Hopf's lemma can be seen as in Corollary 3.2.3.

Chapter 2

Groups of Heisenberg and Iwasawa types explicit solutions to the Yamabe equation

2.1 Introduction

In this section we consider a special class of Carnot groups, those so-called of Heisenberg type. Such groups were introduced by Kaplan [105] and have been subsequently intensively studied by several authors, with [49] and [50] being of crucial importance for the problems considered in this book. We shall construct a family of explicit entire solutions to the Yamabe equation (1.6) on such groups. The second part of the section consists of proving various properties of the Kelvin transform. Such properties are particularly far reaching in the context of Iwasawa groups, where we show that the Kelvin transform is an isometry between the spaces $\overset{o}{\mathcal{D}}^{1,2}(\Omega)$ and $\overset{o}{\mathcal{D}}^{1,2}(\Omega^*)$, where Ω^* denotes the image of Ω under a suitably defined inversion. This will be a useful fact when we consider equations on unbounded domains. In particular, we can obtain the precise asymptotic behavior at infinity of finite energy solutions to the Yamabe equation. It is worth mentioning that this behavior can be obtained also without the use of the Kelvin transform, see [119], which is useful when studying finite energy solutions on non-compact Riemannian and sub-Riemannian manifolds, see also [167].

2.2 Groups of Heisenberg and Iwasawa types

In this section we give the definitions of groups of Heisenberg and Iwasawa types. In addition, we shall include a background, albeit incomplete, which nevertheless points to the importance of these groups in analysis and geometry. In this section it will be convenient to use the established notation N for groups of Heisenberg or Iwasawa type rather than G as we do in the rest of the book.

Let $\mathfrak{n}$ be a 2-step nilpotent Lie algebra equipped with a scalar product $< .,. >$ for which $\mathfrak{n} = V_1 \oplus V_2$-an orthogonal direct sum, V_2 is the center of $\mathfrak{n}$. Consider the map $J : V_2 \to End(V_1)$ defined by

$$< J(\xi_2)\xi_1', \xi_1'' > \; = \; < \xi_2, [\xi_1', \xi_1''] >, \quad \text{for } \xi_2 \in V_2 \text{ and } \xi_1', \xi_1'' \in V_1. \quad (2.1)$$

By definition we have that $J(\xi_2)$ is skew-symmetric. Adding the additional condition that it is actually an almost complex structure on V_1 when ξ_2 is of unit length [105] brings about the next definition.

Definition 2.2.1.

a) A 2-step nilpotent Lie algebra $\mathfrak{n}$ is said to be of *Heisenberg type* if for every $\xi_2 \in V_2$, with $|\xi_2| = 1$, the map $J(\xi_2) : V_1 \to V_1$ is orthogonal.

b) A simply connected connected Lie group N is called of Heisenberg type (or H-type) if its Lie algebra $\mathfrak{n}$ is of Heisenberg type.

We stress that the scalar product needs to be compatible with the linear operator J in order to have a group of *H-type*, see the discussion below concerning the "standard" Heisenberg groups (2.7). We shall continue to use the exponential coordinates and regard $N = exp \, \mathfrak{n}$, so that the product of two elements of N is

$$(\xi_1, \xi_2) \cdot (\xi_1', \xi_2') = (\xi_1 + \xi_1', \xi_2 + \xi_2' + \frac{1}{2}[\xi_1, \xi_1']), \quad (2.2)$$

taking into account the Baker-Campbell-Hausdorff formula (1.7). In [105] Kaplan found the explicit form of the fundamental solution of the sub-Laplacian (1.16) on every group of H-type. We note that when working with agroup of Heisenberg type we shall consider the sub-Laplacian the operator

$$\mathcal{L} = - \sum_{j=1}^{m} X_j^* X_j \; = \; \sum_{j=1}^{m} X_j^2, \quad (2.3)$$

where the vector fields X_j, $j = 1, \ldots, m$ are an orthonormal basis of V_1. In a subsequent paper [106], exploiting the Clifford module structure of V_1, he studied the left invariant Riemannian structure and the associated compact nil-manifolds induced by the fixed scalar product on $\mathfrak{n}$. In particular, Kaplan showed that the group of isometries of N is the semidirect product product $A(N) \ltimes N$ (with N acting by left translations). Here, $A(N)$ denotes the group of automorphisms of N (or $\mathfrak{n}$) that preserve the left-invariant metric, see also [143]. In particular $A(N)$ preserves the decomposition $V_1 \oplus V_2$.

Remark 2.2.2. In regards to the above group of isometries, the reader might want to recall the construction of the Euclidean group $I(\mathbb{R}^k) = O(k) \ltimes \mathbb{R}^k$ with $I_o(\mathbb{R}^k) = SO(k) \ltimes \mathbb{R}^n$ the connected component of the identity. The semidirect product becomes apparent by recalling that the product of two isometries $(A, b) \in O(k) \times \mathbb{R}^n$ and $(A', b') \in O(n) \times \mathbb{R}^n$ representing, respectively, $y \mapsto Ay + b$ and $x \mapsto A'x + b' = y$, is given by $(A, b) \circ (A', b') = (AA', b + Ab')$ in agreement with the composition rule. The Euclidean space R^k can be regarded as the degenerate case of groups of Heisenberg type when V_1 is trivial and the groups is Abelian. We shall not consider explicitly this case in what is to follow since some of the formulas need to be adjusted for this special case.

The definition of J and the orthogonality assumption respectively imply

$$< J(\xi_2)\xi_1, \xi_1 > = 0, \qquad |J(\xi_2)\xi_1| = |\xi_2| \, |\xi_1|. \tag{2.4}$$

It is easy to see that we have in addition

$$J(\xi_2)^2 = -|\xi|^2 \, Id_{V_1}, \qquad < J(\xi_2)\xi_1, J(\xi_2')\xi_1 > = < \xi_2, \xi_2' > |\xi_1|^2,$$
$$J(\xi_2)J(\xi_2') + J(\xi_2')J(\xi_2) = -2 < \xi_2, \xi_2' > Id_{V_1}, \tag{2.5}$$

for any $\xi_1 \in V_1$ and $\xi_2, \xi_2' \in V_2$.

Shortly after the introduction of groups of Heisenberg type Kaplan and Putz [108], see also [[115], Proposition 1.1], observed that the nilpotent part N in the Iwasawa decomposition $\boldsymbol{G} = NAK$ of every semisimple Lie group $\boldsymbol{G}$ of real rank one is of Heisenberg type. We shall refer to such a group as *Iwasawa group* and call the corresponding Lie algebra *Iwasawa algebra*. Specifically, by the Iwasawa decomposition we have $\mathfrak{g} = \mathfrak{n} \oplus \mathfrak{a} \oplus \mathfrak{k}$ with $\mathfrak{n} = \mathfrak{g}_{-\alpha} \oplus \mathfrak{g}_{-2\alpha}$ a direct sum of the (restricted) negative root spaces. Letting $V_1 = \mathfrak{g}_{-\alpha}$, $V_2 = \mathfrak{g}_{-2\alpha}$, a scalar product turning N into an H-type group and the almost complex structures J are given by

$$< \xi, \xi' > = -\frac{1}{m + 4k} B(\xi, \theta\xi'), \qquad \xi, \xi' \in \mathfrak{n},$$
$$J(\xi_2) = ad(\xi_2)\theta, \qquad \xi_2 \in V_2, \tag{2.6}$$

where B and θ are the killing form and the Cartan involution, and $m = \dim \mathfrak{g}_{-\alpha}$, $k = \dim \mathfrak{g}_{-2\alpha}$ are the dimensions of the negative (restricted) root spaces. Very often the notation $\mathfrak{n}$ is used for the space generated by the positive roots $\mathfrak{g}_\alpha \oplus \mathfrak{g}_{2\alpha}$ and $\bar{\mathfrak{n}} = \theta(\mathfrak{n})$ is the standard notation for the space we defined above. From a geometrical point of view, the above Iwasawa groups can be seen as the nilpotent part in the Iwasawa decomposition of the isometry group of the non-compact symmetric spaces M of rank one.

Such a space can be expressed as a homogeneous space G/K where G is the identity component of the isometry group of M, i.e., one of the simple Lorentz groups $SO_o(n,1)$, $SU(n,1)$, $Sp(n,1)$ or $F_{4(-20)}$, and K is a maximal compact subgroup of G, see [88]. Thus, $K = SO(n)$, $S(U(n)U(1))$, $Sp(n)Sp(1)$, or $Spin(9)$, respectively, see for example [171] or [88]. In this way M becomes one of the hyperbolic spaces over the real, complex, quaternion or Cayley (octonion) numbers, respectively. Writing $G = NAK$ and letting $S = NA$, A-one-dimensional Abelian subalgebra, we have that S is a closed subgroup of G, which is isometric with the hyperbolic space M, thus giving the corresponding hyperbolic space a Lie group structure. The nilpotent part N is isometrically isomorphic to $\mathbb{R}^n$ in the degenerate case when the Iwasawa group is Abelian, see Remark 2.2.2, or to one of the Heisenberg groups $G(\mathbb{K}) = \mathbb{K}^n \times \text{Im}\,\mathbb{K}$ with the group law given by

$$(q_o, \omega_o) \circ (q, \omega) = (q_o + q, \omega + \omega_o + 2\,\text{Im}\,q_o\bar{q}), \qquad (2.7)$$

where q, $q_o \in \mathbb{K}^n$ and $\omega, \omega_o \in \text{Im}\,\mathbb{K}$. Here, $\mathbb{K}$ denotes one of the real division algebras: the real numbers $\mathbb{R}$, the complex numbers $\mathbb{C}$, the quaternions $\mathbb{H}$, or the octonions $\mathbb{O}$, see Section 4.3.4 for further details in the quaternion case. In particular, in the non-Euclidean case the Lie algebra $\mathfrak{n}$ of N has center of dimension $\dim V_2 = 1, 3$, or 7. On the other hand, there exist non-Euclidean H-type groups with centers of arbitrary dimensions as shown in [105], hence the Iwasawa groups are a proper subset of the set of all H-type groups.

Iwasawa groups are distinguished also by the properties of the sphere product $S_1(R_1) \times S_2(R_2)$ where $S_j(R_j)$ is the sphere of radius R_j in V_j, $j = 1, 2$. In fact, for a group of Iwasawa type the Kostant double-transitivity theorem shows that the action of $A(N)$ is transitive, where as before $A(N)$ stands for the orthogonal automorphisms of N, see [[50], Proposition 6.1]. This fact points to the importance of the bi-radial or cylindrically symmetric functions, see Definition 3.1.1. Notice that both the fundamental solution (2.13) of the sub-Laplacian and the solution (2.48) of the Yamabe equation posses such symmetry.

Given the ubiquitous role of the Heisenberg group $G(\mathbb{C})$ in analysis, we give some explicit formulas in this special setting. These formulas will also be made explicit in Part 2 for quaternionic contact structures in which case the quaternionic Heisenberg group will play the role of the flat model space, see 4.3.4. $G(\mathbb{C})$ arises as the nilpotent part in the Iwasawa decomposition of the complex hyperbolic space. Specifically, $G(\mathbb{C})$ is a Lie group whose underlying manifold is $\mathbb{C}^n \times \mathbb{R}$ with group law

$$(z,t)(z',t') = (z + z', t + t' + 2Im(z \cdot \bar{z}')), \qquad (2.8)$$

where for $z, z' \in \mathbb{C}^n$ we have let $z \cdot z' = \sum_{j=1}^n z_j z_j'$. In real coordinates a basis for the Lie algebra of left-invariant vector fields on $G(\mathbb{C})$ is given by

$$X_j = \frac{\partial}{\partial x_j} + 2y_j \frac{\partial}{\partial t}, \quad X_{n+j} \equiv Y_j = \frac{\partial}{\partial y_j} - 2x_j \frac{\partial}{\partial t}, \quad \frac{\partial}{\partial t}, \quad j = 1, ..., n. \quad (2.9)$$

Here, we have identified $z = x + iy \in \mathbb{C}^n$, with the real vector $(x, y) \in \mathbb{R}^{2n}$. Since $[X_j, Y_k] = -4\delta_{jk} \frac{\partial}{\partial t}$, the Lie algebra is generated by the system $X = \{X_1, ..., X_{2n}\}$. The relative sub-Laplacian $\mathcal{L} = \sum_{j=1}^{2n} X_j^2$ is the real part of the Kohn complex Laplacian. In this case the exponential map is the identity and, as for any group of step two, the dilations are the parabolic ones $\delta_\lambda(z, t) = (\lambda z, \lambda^2 t)$. The corresponding homogeneous dimension is $Q = 2n + 2$. A little care has to be taken when defining the scalar product which turns $G(\mathbb{C})$ into a group of Heisenberg type. For example, the standard inner product of $\mathbb{C}^n \times \mathbb{R}$, i.e., the inner product in which the basis of left invariant vector fields given in (2.9) is an orthonormal basis will not make the Heisenberg group $G(\mathbb{C})$ a group of Heisenberg type. For example, if this were the case, we would have $J(\frac{\partial}{\partial t})X_i = \sum_{j=1}^{2n} < J(\frac{\partial}{\partial t})X_i, X_j > = -4Y_i$, $i = 1, \dots, n$, which contradicts the orthogonality of $J(\frac{\partial}{\partial t})$. However, this calculation shows that an orthonormal basis of an H-type compatible metric is given by, $j = 1, ..., n$,

$$X_j = \frac{\partial}{\partial x_j} + 2y_j \frac{\partial}{\partial t}, \quad X_{n+j} \equiv Y_j = \frac{\partial}{\partial y_j} - 2x_j \frac{\partial}{\partial t}, \quad T = \frac{1}{4} \frac{\partial}{\partial t}. \quad (2.10)$$

To compute the corresponding homogeneous gauge (1.12) we let $X = \sum_{j=1}^{2n} a_j X_j + bT$ be an element of the Lie algebra and use the orthonormality condition to derive the standard formula

$$|X|^4 = \left(\sum_{j=1}^{2n} a_j^2 \right)^2 + 16b^2, \text{ i.e., } |(z,t)| = (|z|^4 + 16t^2)^{1/4}, \quad (2.11)$$

the latter equation written for the corresponding gauge on the group using $|z| = \left(\sum_{j=1}^n (x_j^2 + y_j^2) \right)^{1/2}$ - the Euclidean norm.

On a group N of Heisenberg type there is a very important homogeneous norm (gauge) given by

$$N(g) = \left(|x(g)|^4 + 16|y(g)|^2 \right)^{1/4}, \quad (2.12)$$

which induces a left-invariant distance, cf. (1.15). Kaplan proved in [105] that in a group of Heisenberg type the fundamental solution Γ of the sub-Laplacian $\mathcal{L}$, see (2.3), is given by the formula

$$\Gamma(g, h) = C_Q \, N(h^{-1}g)^{-(Q-2)}, \qquad g, h \in N, g \neq h, \quad (2.13)$$

where C_Q is a suitable constant. Equation (2.13) will play a key role in Definition 2.3.2 below.

Remark 2.2.3. It is known that the distance induced by the gauge (2.12) is the Gromov limit of a one parameter family of Riemannian metrics on the group N [115], see also [21] and [36].

It should be noted that frequently when working exclusively with the Heisenberg group G ($\mathbb{C}$) one uses the gauge

$$|(z,t)| = (|z|^4 + t^2)^{1/4}, \tag{2.14}$$

which may be the cause for a difference in some constants. Similar convention is also used in the quaternionic setting.

Motivated by the way the Iwasawa type groups appear as "boundaries" of the hyperbolic spaces, Damek [56] introduced a generalization of the hyperbolic spaces by starting with a group N of H-type and taking a semidirect product with a one dimensional Abelian group. Specifically, consider the multiplicative group $A = \mathbb{R}^+$ acting on an H-type group by dilations given in exponential coordinates by the formula $\delta_a(\xi_1, \xi_2) = (a^{1/2}\xi_1, a\xi_2)$ and define $S = NA$ as the corresponding semidirect product. Thus, the Lie algebra of S is $\mathfrak{s} = V_1 \oplus V_2 \oplus \mathfrak{a}$, $\mathfrak{n} = V_1 \oplus V_2$, with the bracket extending the one on $\mathfrak{n}$ by adding the rules

$$[\zeta, \xi_1] = \frac{1}{2}\xi_1, \quad [\zeta, \xi_2] = \xi_2 \quad \xi_i \in V_i, \tag{2.15}$$

where ζ is a unit vector in $\mathfrak{a}$, so that S is the connected simply connected Lie group with Lie algebra $\mathfrak{s}$. In the coordinates $(\xi_1, \xi_2, a) = \exp(\xi_1 + \xi_2)\exp(\log a\zeta)$, $a > 0$, which parameterize $S = exp\,\mathfrak{s}$, the product rule of S is given by the formula

$$(\xi_1, \xi_2, a) \cdot (\xi_1', \xi_2', a') = (\xi_1 + a^{1/2}\xi_1', \, \xi_2 + a\xi_2' + \frac{1}{2}a^{1/2}[\xi_1, \xi_1'], \, aa'), \tag{2.16}$$

for all $(\xi_1, \xi_2, a), (\xi_1', \xi_2', a') \in \mathfrak{n} \times \mathbb{R}^+$. Notice that S is a solvable group. We equip the Lie algebra $\mathfrak{s}$ with the inner product

$$< (\xi_1, \xi_2, a), (\tilde{\xi}_1, \tilde{\xi}_2, \tilde{a}) > \, = \, < (\xi_1, \xi_2), (\tilde{\xi}_1, \tilde{\xi}_2) > + a\tilde{a} \tag{2.17}$$

using the fixed inner product on $\mathfrak{n}$ and then define a corresponding translation invariant Riemannian metric on S. The main result of [56] is then that the group of isometries $Isom(S)$ of S is as small as it can be and equals $A(S) \ltimes S$ with S acting by left translations, unless N is one of the Heisenberg groups (2.7), i.e., S is one of the classical hyperbolic spaces.

Here, $A(S)$ denotes the group of automorphisms of S (or $\mathfrak{s}$) that preserve the left-invariant metric on S. The spaces constructed in this manner became known as Damek-Ricci spaces, see [22] for more details. It is worth recalling [57] where it is shown that the just described solvable extension of H-type groups, which are not of Iwasawa type, provide noncompact counterexamples to a conjecture of Lichnerowicz, which asserted that harmonic Riemannian spaces must be rank one symmetric spaces.

The Heisenberg type groups allowed for the generalization of many important concepts in harmonic analysis and geometry, see [108], [109], [115], [58] and the references therein, in addition to the above cited papers. Another milestone was achieved in [49], which allowed for avoiding the classification rank one symmetric spaces and the heavy machinery of the semisimple Lie group theory, when studying the non-compact symmetric spaces of real rank one. Specifically, in [49] the authors considered the H-type algebras satisfying the so called J^2 condition defined in [49], see also [50].

Definition 2.2.4. We say that the H-type algebra $\mathfrak{g}$ satisfies the J^2 condition if for every $\xi_2, \xi_2' \in V_2$ which are orthogonal to each other, $< \xi_2, \xi_2' > = 0$, there exists $\xi_2'' \in V_2$ such that

$$J(\xi_2)J(\xi_2') = J(\xi_2''). \tag{2.18}$$

The key result here is the following Theorem of [49], see also [47], which can be used to show that if N is an H-type group, then the Riemannian space $S = NA$ is symmetric iff the Lie algebra $\mathfrak{n}$ of N satisfies the J^2 condition, see [[49], Theorem 6.1].

Theorem 2.2.5. *If* $\mathfrak{n}$ *is an H-type algebra satisfying the* J^2*-condition, then* $\mathfrak{n}$ *is an Iwasawa type algebra. In other words, the H-type groups* N *whose Lie algebras satisfy the* J^2 *condition are precisely the groups that arise as the nilpotent component in the Iwasawa decomposition of a semisimple Lie group of real rank one.*

This fundamental result has many consequences in allowing a unified proof of some classical results on symmetric spaces, in addition to some beautiful properties of extensions of the classical Cayley transform, inversion and Kelvin transform, which are of a particular importance for our goals.

2.3   The Cayley transform, inversion and Kelvin transform

2.3.1   *The Cayley transform*

In this section we focus on the Cayley transform, of which we shall make use later in Example 5.2.1 in the case of the quaternionic Heisenberg group. Starting from an H-type group, its solvable extension S has the following realizations, [58], [49] and [50].

First, it can be viewed as a "Siegel domain" or an upper-half plane model of the hyperbolic space

$$D = \{p = (\xi_1, \xi_2, a) \in \mathfrak{s} = V_1 \oplus V_2 \oplus \mathfrak{a} : a > \frac{1}{4}|\xi_1|^2\}. \tag{2.19}$$

Consider the map $\Theta : S \to S$,

$$\Theta(\xi_1, \xi_2, a) = (\xi_1, \xi_2, a + \frac{1}{4}|\xi_1|^2), \tag{2.20}$$

which is injective map of S into itself. Here we use a to denote the element $a\zeta \in A$, ζ defined after (2.15), and we regard D as a subset of S using the exponential coordinates. Thus, the group S acts simply transitively on D by conjugating left multiplication in the group S by Θ, $s \cdot p = \Theta s \cdot (\Theta^{-1}p)$ for $s \in S$ and $p \in D$, while N acts simply transitively on the level sets of $h = a - \frac{1}{4}|\xi_1|^2$. In particular, we can define an invariant metric on D by pulling via Θ the left-invariant metric (2.17) of S to D, thus making Θ an isometry, cf. [[50], (3.3)]. It is very instructive to see this explicitly in the case of the Heisenberg group $G(\mathbb{C})$ for which the reader can consult [[153], Chapter XII].

Second, there is the "ball" model of S,

$$B = \{(\xi_1, \xi_2, a) \in \mathfrak{s} = V_1 \oplus V_2 \oplus \mathfrak{a} : |\xi_1|^2 + |\xi_2|^2 + a^2 < 1\}, \tag{2.21}$$

equipped with the metric obtained from D via the inverse of the so-called *Cayley transform* $\mathcal{C} : B \to D$ defined by $\mathcal{C}(\xi_1, \xi_2, a) = (\xi_1', \xi_2', a')$, where

$$\xi_1' = \frac{2}{(1-a)^2 + |\xi_2|^2} \left((1-a)\xi_1 + J(\xi_2)\xi_1\right),$$

$$\xi_2' = \frac{2}{(1-a)^2 + |\xi_2|^2}\xi_2, \qquad a' = \frac{1 - a^2 - |\xi_2|^2}{(1-a)^2 + |\xi_2|^2}. \tag{2.22}$$

The inverse map $\mathcal{C}^{-1} : D \to B$ is given by $\mathcal{C}^{-1}(\xi_1', \xi_2', a') = (\xi_1, \xi_2, a)$, where

$$\xi_1 = \frac{2}{(1+a')^2 + |\xi_2'|^2} \left((1+a')\xi_1' - J(\xi_2')\xi_1'\right),$$

$$\xi_2 = \frac{2}{(1+a')^2 + |\xi_2'|^2}\xi_2', \qquad a = \frac{-1 + a'^2 - |\xi_2'|^2}{(1+a')^2 + |\xi_2'|^2}. \tag{2.23}$$

The proof of the above formulas is elementary. For example, we can see that $\mathcal{C}(\xi_1, \xi_2, a) \in D$, i.e., $a' > \frac{1}{4}|\xi_1'|^2$ from the computation

$$a' = \frac{1 - a^2 - |\xi_2|^2}{(1 - a)^2 + |\xi_2|^2} > \frac{|\xi_1|^2}{(1 - a)^2 + |\xi_2|^2},$$

$$|\xi_1'|^2 = \frac{4}{\left((1 - a)^2 + |\xi_2|^2\right)^2} |(1 - a)\xi_1 + J(\xi_2)\xi_1|^2 \tag{2.24}$$

$$= \frac{4\left((1 - a)^2|\xi_1|^2 + |\xi_2|^2|\xi_1|^2\right)}{\left((1 - a)^2 + |\xi_2|^2\right)^2} = \frac{4|\xi_1|^2}{(1 - a)^2 + |\xi_2|^2},$$

using equations (2.4) in the next to last equality. For other versions of the Cayley transform see [[63], Chapter X]. The Jacobian of $\mathcal{C}$ and its determinant were computed in [58]. The latter is given by the formula

$$\det \mathcal{C}'(\xi_1, \xi_2, a) = 2^{m+k+1} \left((1 - a)^2 + |\xi_2|^2\right)^{-(m+2k+2)/2}, \tag{2.25}$$

where, as before, $m = \dim V_1$, $k = \dim V_2$.

It is very important and we shall make use of the fact that the Cayley transform can be extended by continuity to a bijection (denoted by the same letter!)

$$C : \partial B \setminus \{(0, 0, 1)\} \to \partial D, \tag{2.26}$$

where $(0, 0, 1)$ (referred to as "ζ" for short) is the point on the sphere where $\xi_1 = \xi_2 = 0$ and the third component is ζ in agreement without notation set after equation (2.20). The boundaries of the ball and Siegel domain models are, respectively,

$$\Sigma \equiv \partial D = \{p = (\xi_1', \xi_2', a') \in \mathfrak{s} = V_1 \oplus V_2 \oplus \mathfrak{a} : a' = \frac{1}{4}|\xi_1'|^2\} \tag{2.27}$$

and

$$\partial B = \{(\xi_1, \xi_2, a) \in \mathfrak{s} = V_1 \oplus V_2 \oplus \mathfrak{a} : |\xi_1|^2 + |\xi_2|^2 + a^2 = 1\}. \tag{2.28}$$

The group of Heisenberg type N can be identified with Σ via the map

$$(\xi_1', \xi_2') \mapsto (\xi_1', \xi_2', \frac{1}{4}|\xi_1'|^2). \tag{2.29}$$

With this identification we obtain the form of the Cayley transform (stereographic projection) identifying the sphere minus the point "ζ" and the H-type group, $\mathcal{C} : \partial B \setminus \{(0, 0, 1)\} \to N$ defined by $\mathcal{C}(\xi_1, \xi_2, a) = (\xi_1', \xi_2')$, where

$$\xi_1' = \frac{2}{(1 - a)^2 + |\xi_2|^2} \left((1 - a)\xi_1 + J(\xi_2)\xi_1\right),$$

$$\xi_2' = \frac{2}{(1 - a)^2 + |\xi_2|^2} \xi_2. \tag{2.30}$$

Later, we shall encounter the "boundary" Cayley transform, also called (generalized) stereographic projection, in the case of the quaternionic Heisenberg group. At that place we shall give some other explicit formulas valid in the quaternionic contact setting, see Section 5.2.1. In particular, we shall see that the Cayley transform is a quaternionic contact conformal transformation. We shall exploit this fact in the solution of the quaternionic contact Yamabe problem on the seven dimensional quaternionic Heisenberg group and sphere Theorem 6.1.2 and in the determination of the best constant in the Folland-Stein embedding inequality on *any* quaternionic Heisenberg group Theorem 6.1.1. However, it is worth noting that, in fact, given any Iwasawa group N the Cayley transform preserves the horizontal space, i.e., the Cayley transform is a multicontact transformation. This fact should allow for the generalization of Theorem 6.1.1 and the earlier result on the Heisenberg group [70] in the unified setting of groups of Iwasawa type. The Cayley transform is also a 1-quasiconformal map [20], see also [13]. The definition of the "horizontal" space in the tangent bundle of the sphere and the distance function on the sphere require a few more details for which we refer to [50] and [20]. Multicontact maps and their rigidity in Carnot groups have been studied recently in [137], [142], [116], [52], [53], [35], [59].

2.3.2 Inversion on groups of Heisenberg type

In [114] Korányi introduced an inversion on the Heisenberg group and used it to define an analogue of the Kelvin transform in such setting. Subsequently, such inversion formula, as well as the Kelvin transform, were generalized in [51] and [49] to all groups of Heisenberg type.

Similarly to the Cayley transform, the inversion on groups of Heisenberg type can be seen as the restriction to the boundary of an inversion transformation on the ball (2.21) or Siegel domain (2.19) models. On the latter space the inversion is defined by the formula

$$\sigma(\xi_1, \xi_2, a) = \frac{1}{|\xi_2|^2 + a^2} \left((-a + J(\xi_2)) \xi_1, -\xi_2, a \right). \tag{2.31}$$

The formula on the ball B is obtained by transporting the above inversion to B via the Cayley transform, i.e., conjugating with $\mathcal{C}$. Remarkably, it was proven in [[49], Theorem 6.1], see also [[50], Theorem 4.1], that σ is an isometry of D if and only if the J^2 condition holds on the Heisenberg type group N. This happens exactly when S is a symmetric space.

We turn to the definition of the inversion as a map on a group of Heisenberg type.

Definition 2.3.1. Let G be a group of Heisenberg type with Lie algebra $\mathfrak{g} = V_1 \oplus V_2$. For $g = \exp(\xi) \in G$, with $\xi = \xi_1 + \xi_2$, the *inversion* $\sigma : G^* \to G^*$, where $G^* = G \setminus \{e\}$ is defined by

$$\sigma(g) = \left(- \left(|x(g)|^2 \, Id_{V_1} + 4J(\xi_2) \right)^{-1} \xi_1, - \frac{\xi_2}{|x(g)|^4 + 16|y(g)|^2} \right), \quad (2.32)$$

where the map J is as in (2.1), and Id_{V_1} denotes the identity map on V_1. One easily verifies that

$$\sigma^2(g) = g, \qquad g \in G^*.$$

As before, see (2.12), $N(g)$ will be the gauge $N(g) = \left(|x(g)|^4 + 16|y(g)|^2 \right)^{1/4}$. Writing $\sigma(g) = \exp(\eta)$, with $\eta = \eta_1 + \eta_2$, for the image of g, we easily obtain from Definition 2.3.1 and (2.4) that

$$|\eta_1| = \frac{|\xi_1|}{N(g)^2}, \qquad \text{and} \qquad |\eta_2| = \frac{|\xi_2|}{N(g)^4} \qquad (2.33)$$

An immediate consequence of (2.33) is the identity

$$N(\sigma(g)) = N(g)^{-1}, \qquad g \in G^*. \qquad (2.34)$$

A direct verification using (2.1) and the definition of the group dilations shows that the inversion anti-commutes with the group dilations (1.9), i.e.,

$$\sigma(\delta_\lambda(g)) = \delta_{\lambda^{-1}}(\sigma(g)), \qquad g \in G^*. \qquad (2.35)$$

We finish this sub-section by writing the formulas for the inversion on $G(\mathbb{C})$, see [114]. Let $A = |z|^2 + it$ so that $A\bar{A} = |(z,t)|^4$, cf. (2.14). The inversion of a point (z, t) is given by

$$(w, \tau) = \sigma(z, t) \stackrel{def}{=} \left(- \frac{z}{\bar{A}}, - \frac{t}{A\bar{A}} \right).$$

The expression of the inversion in the real variables is

$$u = - \frac{|z|^2 x - ty}{|z|^4 + t^2}, \qquad v = - \frac{|z|^2 y + tx}{|z|^4 + t^2}, \qquad \tau = - \frac{t}{|z|^4 + t^2}, \quad (2.36)$$

where $w = u + iv$.

2.3.3 The Kelvin transform

The purpose of this section is to recall the relevant definitions and establish some more properties of the Kelvin transform, [51] and [49]. Such properties are particularly far reaching in the context of Iwasawa groups, where we show that the Kelvin transform is an isometry between the spaces $\overset{o}{\mathcal{D}}{}^{1,2}(\Omega)$ and $\overset{o}{\mathcal{D}}{}^{1,2}(\Omega^*)$, where Ω^* denotes the image of Ω under the inversion (2.32) on the group. This will be a useful fact when considering the Yamabe equation on unbounded domains.

Definition 2.3.2. Let G be a group of Heisenberg type, and consider a function u on G. The *Kelvin transform* of u is defined by the equation

$$u^*(g) \; = \; N(g)^{-(Q-2)}\, u(\sigma(g)), \qquad g \in G^*, \qquad (2.37)$$

where σ is the inversion on the group and $N(g)$ is the gauge (2.12).

When G is a group of Iwasawa type, then it was proved in [49] that the inversion and the Kelvin transform possess various properties generalizing those of the well-known Kelvin transform of functions defined on the Euclidean space. In the following theorem we collect the two which will be used in this book.

Theorem 2.3.3 (see [49]). *Let G be a group of Iwasawa type and $\mathcal{L}$ the sub-Laplacian operator (2.3). The Jacobian of the inversion is given by*

$$d(H \circ \sigma)(g) \; = \; N(g)^{-2Q}\, dH(g), \qquad\qquad g \in G^*$$

using the Haar measure dH on the group, cf. (1.2.1). The Kelvin transform u^* of a function satisfies the equation

$$\mathcal{L}u^*(g) \; = \; N(g)^{-(Q+2)}(\mathcal{L}u)(\sigma(g)), \qquad\qquad g \in G^*.$$

Remark 2.3.4. We shall denote by Ω^* the image of a generic domain Ω under the inversion σ. We stress that, since we have chosen not to define the inversion of the point at infinity, in the case in which Ω is a neighborhood of ∞, by which we mean that there exists a ball $B(e, R)$ such that $(G \backslash \overline{B}(e, R)) \subset \Omega$, then Ω^* is a punctured neighborhood of the identity, i.e., $\Omega^* = D \setminus \{e\}$, for an open set D such that $e \in D$. The reader should keep this point in mind for the proof of the next result, as well as for the results in the following sections. The following theorem is a consequence of the conformal properties of the inversion and of the Kelvin transform. Such result will be used in the next section in combination with the conformal invariance of the Yamabe type equation expressed by Lemma 2.3.6.

Theorem 2.3.5. *The Kelvin transform is an isometry between* $\overset{o}{\mathcal{D}}{}^{1,2}(\Omega)$ *and* $\overset{o}{\mathcal{D}}{}^{1,2}(\Omega^*)$.

Proof. Let $u, v \in \overset{o}{\mathcal{D}}{}^{1,2}(\Omega)$ and $u^*, v^* \in \overset{o}{\mathcal{D}}{}^{1,2}(\Omega^*)$ be their Kelvin transforms. We begin by observing that thanks to Theorem 2.3.3 and (2.34)

$$\int_{\Omega^*} (u^*(g'))^{2^*} dH(g') = \int_{\Omega^*} \left[N(g')^{-(Q-2)} u(\sigma(g')) \right]^{2Q/(Q-2)} dH(g')$$

$$= \int_{\Omega} \left[N(\sigma(g))^{-(Q-2)} u(g) \right]^{2Q/(Q-2)} N(g)^{-2Q} dH(g) = \int_{\Omega} u(g)^{2^*} dH(g).$$

We want to show next that

$$\int_{\Omega} <Xu(g), Xv(g)> dH(g) = \int_{\Omega^*} <Xu^*(g'), Xv^*(g')> dH(g').$$
(2.38)

By density it suffices to assume that $u, v \in C_o^\infty(\Omega)$. An integration by parts shows that (2.38) is equivalent to

$$\int_{\Omega} u(g)\, \mathcal{L}v(g)\, dH(g) = \int_{\Omega^*} u^*(g')\, \mathcal{L}v^*(g')\, dH(g').$$

Using again Theorem 2.3.3 and (2.34) we obtain

$$\int_{\Omega^*} u^*(g')\, \mathcal{L}v^*(g')\, dH(g')$$

$$= \int_{\Omega^*} N(g')^{-(Q-2)}\, u(\sigma(g'))\, N(g')^{-(Q+2)}\, (\mathcal{L}v)(\sigma(g'))\, dH(g')$$

$$= \int_{\Omega} u(g)\, \mathcal{L}v(g)\, N(g)^{(Q-2)}\, N(g)^{(Q+2)}\, N(g)^{-2Q}\, dH(g)$$

$$= \int_{\Omega} u(g)\, \mathcal{L}v(g)\, dH(g).$$

This completes the proof. $\qquad\qquad\square$

Next we are going to show that the Yamabe equation is invariant under the Kelvin transform in a sense to be made precise in Theorem 2.3.7. Let G be an Iwasawa group and Ω^* be an unbounded open set. By Ω we denote the image of the open set Ω^* under the inversion with center at the identity e. We recall Remark 2.3.4. We also note that since problem

(2.42) is translation invariant we can by left-translation send Ω^* to another conveniently chosen unbounded domain. If the complement of Ω^* contains a ball we can thus suppose from the beginning that such ball is centered at the group identity e. Furthermore, by a simple rescaling we can without restriction assume that the radius of the ball is one. We start with a simple, yet crucial, lemma.

Lemma 2.3.6. *Let u be a solution of*

$$\begin{cases} \mathcal{L}u = -u^p \\ u \in \overset{o}{\mathcal{D}}{}^{1,2}(\Omega), \quad u \geq 0, \end{cases} \tag{2.39}$$

and denote by u^* its Kelvin transform. Then u^* satisfies

$$\mathcal{L}u^*(g) = -N(g)^{p(Q-2)-(Q+2)} u^*(g)^p \qquad g \in \Omega^*. \tag{2.40}$$

In particular, when $p = \frac{Q+2}{Q-2}$, if u is a solution in Ω of the equation

$$\mathcal{L}u = -(u)^{(Q+2)/(Q-2)}, \quad u \in \overset{o}{\mathcal{D}}{}^{1,2}(\Omega), \quad u \geq 0, \tag{2.41}$$

then u^* is a solution in Ω^* of the equation

$$\mathcal{L}u^* = -(u^*)^{(Q+2)/(Q-2)}, \quad u^* \in \overset{o}{\mathcal{D}}{}^{1,2}(\Omega^*), \quad u^* \geq 0. \tag{2.42}$$

Proof. Let u be a solution to (2.39). From Theorem 2.3.5 we know $u^* \in \overset{o}{\mathcal{D}}{}^{1,2}(\Omega^*)$. Consider an arbitrary function $\psi \in C_o^\infty(\Omega^*)$, then we can write $\psi = \phi^*$, for some $\phi \in C_o^\infty(\Omega)$. Integrating by parts and applying Theorem 2.3.3 gives

$$\int_{\Omega^*} <Xu^*(g'), X\psi(g')> dH(g') = \int_{\Omega^*} <Xu^*(g'), X\phi^*(g')> dH(g')$$

$$= -\int_{\Omega^*} u^*(g')\mathcal{L}\phi^*(g')dH(g') = -\int_{\Omega^*} N(g')^{-2Q} u(\sigma(g'))\mathcal{L}\phi^*(\sigma(g'))dH(g')$$

$$= -\int_{\Omega} u(g)\mathcal{L}\phi(g)dH(g) = \int_{\Omega} <Xu(g), X\phi(g)> dH(g)$$

$$= \int_{\Omega} u(g)^p\phi(g)dH(g),$$

where in the last equality we have used the fact that u is a solution to (2.39). We now make the change of variable $g = \sigma(g')$, $g' \in \Omega^*$, and use Theorem 2.3.3 again to obtain

$$\int_{\Omega} u(g)^p\phi(g)dH(g) = \int_{\Omega^*} u(\sigma(g'))^p\phi(\sigma(g'))N(g')^{-2Q}dH(g')$$

$$= \int_{\Omega^*} u^*(\sigma(g'))^p\phi^*(\sigma(g'))N(g')^{(Q-2)p-(Q+2)}dH(g').$$

In conclusion we have found

$$\int_{\Omega^*} < Xu^*(g'), X\psi(g') > dH(g')$$
$$= \int_{\Omega^*} u^*(\sigma(g'))^p \phi^*(\sigma(g')) N(g')^{(Q-2)p-(Q+2)} dH(g').$$

By the arbitrariness of $\psi \in C_o^\infty(\Omega^*)$, (2.40) follows. □

In the following theorem we show that if u^* is a solution to (2.42) in a neighborhood of infinity (see Remark 2.3.4), then the Kelvin transform of u^* has a removable singularity at the group identity e.

Theorem 2.3.7. *Let G be an Iwasawa group. Suppose that u^* is a solution of (2.42) in Ω^*, with Ω^* a neighborhood of infinity. Let u be the Kelvin transform of u^* defined in Ω, then the group identity e is a removable singularity, i.e., u can be extended as a smooth function in a neighborhood of e where the equation is satisfied.*

Proof. Due to the assumptions on Ω^* we can write $\Omega = D \backslash \{e\}$, where D is a bounded open set containing e. Theorem 2.3.5 implies that $u \in \overset{o}{\mathcal{D}}{}^{1,2}(\Omega)$, moreover from Lemma 2.3.6 (with the roles of u and u^* reversed) we know that u satisfies (2.41) in Ω, hence for every $\psi \in C_o^\infty(D \backslash \{e\})$ one has

$$\int_D < Xu, X\psi > dH = \int_D u^{2^*-1} \psi dH. \tag{2.43}$$

According to [[37], Proposition 6.1], we can find a sequence of functions $\zeta_k \in C^\infty(D \backslash \{e\})$ with $e \notin \text{supp}\,\zeta_k$, such that, $0 \le \zeta_k \le 1$, $\zeta_k(g) \to 1$ for every $g \in D \backslash \{e\}$ and for which

$$\int_D |X\zeta_k|^2 dH \to 0, \tag{2.44}$$

as $k \to \infty$. We fix $\phi \in C_o^\infty(D)$ arbitrarily. For every $k \in \mathbb{N}$ one has $\phi\zeta_k \in C_o^\infty(D \backslash \{e\})$, and therefore we obtain from (2.43)

$$\int_D u^{2^*-1} \phi\zeta_k dH = \int_D < Xu, X(\phi\zeta_k) > dH$$
$$= \int_D \zeta_k < Xu, X\phi > dH + \int_D \phi < Xu, X\zeta_k > dH.$$

Since $u \in \overset{o}{\mathcal{D}}^{1,2}(\Omega) \subset \mathcal{D}^{1,2}(D)$ we can apply Lebesgue dominated convergence theorem to conclude, using (2.44),

$$\int_D < Xu, X\phi > dH = \int_D u^{2^*-1} \phi dH. \tag{2.45}$$

The arbitrariness of $\phi \in C_o^\infty(D)$ shows that the identity is a removable singularity. This completes the proof.   □

2.4   Explicit entire solutions of the Yamabe equation on groups of Heisenberg type

When u is a solution of the Yamabe equation on G,

$$\mathcal{L}u = -u^{2^*-1}, \tag{2.46}$$

then we say that u is an *entire solution* of the Yamabe equation. Kaplan and Putz [[107], Proposition 2] are the first to find an explicit entire solution of the Yamabe equation on groups of Iwasawa type, even though their result seems to have been forgotten. Jerison and Lee [103] made the deep discovery that, up to group translations and dilations, a suitable multiple of the function

$$u(z,t) = ((1+|z|^2)^2 + t^2)^{-(Q-2)/4}, \tag{2.47}$$

is the only positive entire solution of (2.46) on the Heisenberg group $G\,(\mathbb{C})$ (2.8). Here, we have denoted with (z,t), $z \in \mathbb{C}^n, t \in \mathbb{R}$, the variable point in $G\,(\mathbb{C})$ and $Q = 2n + 2$ is, as usual, the homogeneous dimension. The goal of this section is to give an explicit family of entire solutions on groups of Heisenberg type by proving the following result of [[76], Theorem 1.1].

Theorem 2.4.1. *Let G be a group of Heisenberg type. For every $\epsilon > 0$ the function*

$$K_\epsilon(g) = \left(\frac{m(Q-2)\epsilon^2}{(\epsilon^2 + |x(g)|^2)^2 + 16|y(g)|^2} \right)^{\frac{Q-2}{4}}, \qquad g \in G, \tag{2.48}$$

is a positive, entire solution of the Yamabe equation (2.46).

The proof requires a few formulas which are stated in the next Lemmas. Let $k = \dim V_2$ and $m = \dim V_1$. The coordinates of the projection ξ_1 in the basis $X_1,\ldots,X_m$ will be denoted by $x_1 = x_1(g),\ldots,x_m = x_m(g)$, i.e.,

$$x_j(g) = < \xi(g), X_j > \qquad j = 1,\ldots,m, \tag{2.49}$$

and we set $x = x(g) = (x_1, \ldots, x_m) \in \mathbb{R}^m$. Similarly, we fix an orthonormal basis $Y_1, \ldots, Y_k$ of the second layer V_2 and define the exponential coordinates in the second layer V_2 of a point $g \in G$ by letting

$$y_i(g) \ = \ < \xi(g), Y_i >, \qquad\qquad i = 1, \ldots, k, \qquad (2.50)$$

and $y = (y_1, \ldots, y_k) \in \mathbb{R}^k$.

A simple consequence of the Baker-Campbell-Hausdorff formula is the following Lemma. This result shows in particular that in a Carnot group the coordinates in the first and second layers of the Lie algebra $\mathfrak{g}$ are $\mathcal{L}$-harmonic, where as before $\mathcal{L}$ is the associated sub-Laplacian.

Lemma 2.4.2. *Let G be a Carnot group. The function $\psi(g) \overset{def}{=} |x(g)|^2$ enjoys the following properties*

$$|X\psi|^2 = 4\psi, \qquad \mathcal{L}\psi = 2m. \qquad (2.51)$$

Proof. Define the functions

$$\psi_j(t) = |\xi_1(g\ exp(tX_j))|^2 \qquad\qquad j = 1, \ldots, m.$$

The Baker-Campbell-Hausdorff formula implies

$$g\ exp(tX_j) = exp(\xi_1(g) + tX_j + \xi_2(g) + \ldots + \xi_r(g) + \frac{1}{2}[\xi_1(g) + \ldots + \xi_r(g), tX_j] + \ldots).$$

From this one immediately sees the equation

$$\psi_j(t) \ = \ |\xi_1(g)|^2 \ + \ 2t < \xi_1(g), X_j > \ + \ t^2, \qquad (2.52)$$

which implies

$$\psi_j'(0) \ = \ 2 < \xi_1(g), X_j > \ = \ 2x_j(g), \qquad \psi_j''(0) = 2. \qquad (2.53)$$

Now, equation (2.53) gives

$$\mathcal{L}\psi = \sum_{j=1}^{m} \psi_j''(0) = 2m,$$

$$|X\psi|^2 = \sum_{j=1}^{m} \psi_j'(0)^2 = 4\sum_{j=1}^{m} < \xi(g), X_j >^2 = 4\sum_{j=1}^{m} |x_j(g)|^2 = 4\psi,$$

which proves the lemma. $\qquad\qquad\qquad\qquad\qquad\qquad\qquad\qquad\square$

Lemma 2.4.3. *Let **G** be a Carnot group. One has*
$$\mathcal{L}x_j = 0, \quad j = 1, \ldots, m, \qquad\qquad \mathcal{L}y_i = 0, \quad i = 1, \ldots, k.$$
From the latter equation we infer, in particular, that the function $g \to |y(g)|^2$ is $\mathcal{L}$-subharmonic and in fact
$$\mathcal{L}(|y|^2) = 2 \sum_{i=1}^{k} |X(y_i)|^2 \geq 0.$$
There exists a constant $C = C(\mathbf{G}) > 0$ such that
$$|X(|y|^2)|^2 \leq C \, |x|^2 \, |y|^2.$$

Proof. Let $g = exp(\xi)$ with $\xi = \xi_1 + \ldots + \xi_r$. For $t \in \mathbb{R}$ the Baker-Campbell-Hausdorff formula and the stratification of $\mathfrak{g}$ give for $l = 1, \ldots, m$
$$x_j(g \exp tX_l) = x_j(g) + t\delta_{jl},$$
$$y_i(g \exp tX_l) = y_i(g) + \frac{t}{2} < [\xi_1, X_l], Y_i > . \tag{2.54}$$
From (2.54) the $\mathcal{L}$-harmonicity of $x_j(g)$ and $y_i(g)$ is obvious. Using (2.54) we now define for $l = 1, \ldots, m$
$$\phi_l(t) = |y(g \exp tX_l)|^2$$
$$= \sum_{i=1}^{k} \left(y_i^2 + t < [\xi_1, X_l], Y_i > y_i + \frac{t^2}{4} < [\xi_1, X_l], Y_i >^2 \right). \tag{2.55}$$
Differentiating with respect to t we find
$$\phi_l'(0) = \sum_{i=1}^{k} < [\xi_1, X_l], Y_i > y_i, \tag{2.56}$$
hence
$$\phi_l'(0)^2 \leq |y|^2 \sum_{i=1}^{k} (< [\xi_1, X_l], Y_i >)^2.$$
Keeping in mind that $\xi_1 = \sum_{j=1}^{m} x_j X_j$ we easily obtain
$$\sum_{i=1}^{k} (< [\xi_1, X_l], Y_i >)^2 \leq |x|^2 \sum_{j=1}^{m} \sum_{i=1}^{k} (< [X_j, X_l], Y_i >)^2.$$
In conclusion
$$|X(|y|^2)|^2 = \sum_{l=1}^{m} \phi_l(0)^2 \leq C \, |x|^2 \, |y|^2,$$
where
$$C = \sum_{j,l=1}^{m} \sum_{i=1}^{k} (< [X_j, X_l], Y_i >)^2.$$

$\square$

Another technical lemma which we shall need gives the following formulas valid on groups of Heisenberg type.

Lemma 2.4.4. *Let G be a group of Heisenberg type. The following formulas hold*

$$\mathcal{L}(|y(g)|^2) = \frac{k}{2} |x(g)|^2 \tag{2.57}$$

$$|X(|y|^2)|^2 = |x|^2 |y|^2 \tag{2.58}$$

$$< X(|x(g)|^2), X(|y(g)|^2) > = 0, \tag{2.59}$$

where $|x|^2 = x_1^2 + \cdots + x_m^2$, $|y|^2 = y_1^2 + \ldots, y_k^2$.

Proof. Recalling (2.55) one sees

$$\phi_l''(0) = \frac{1}{2} \sum_{i=1}^{k} (< Y_i, [\xi_1, X_l] >)^2 = \frac{1}{2} \sum_{i=1}^{k} (< J(Y_i)\xi_1, X_l >)^2.$$

This implies in view of (2.1), (2.4)

$$\mathcal{L}(|y|^2) = \sum_{l=1}^{m} \phi_l''(0) = \frac{1}{2} \sum_{l=1}^{m} \sum_{i=1}^{k} (< J(Y_i)\xi_1, X_l >)^2$$

$$= \frac{1}{2} \sum_{i=1}^{k} |J(Y_i)\xi_1|^2 = \frac{k}{2} |x|^2.$$

From (2.56) one has

$$\phi_l'(0) = \sum_{i=1}^{k} < [\xi_1, X_l], Y_i >< \xi_2, Y_i > = < \xi_2, [\xi_1, X_l] > . \tag{2.60}$$

Using (2.4) we obtain from the latter equality

$$|X(|y|^2)|^2 = \sum_{l=1}^{m} \phi_l'(0)^2 = \sum_{l=1}^{m} (< J(\xi_2)\xi_1, X_l >)^2 = |J(\xi_2)\xi_1|^2 = |x|^2 |y|^2.$$

Finally, (2.52), (2.60), (2.1) and (2.4) imply

$$< X(|x|^2), X(|y|^2) > = 2 \sum_{l=1}^{m} < \xi_1, X_l >< \xi_2, [\xi_1, X_l] >$$

$$= 2 < J(\xi_2)\xi_1, \xi_1 > = 0.$$

This completes the proof. $\qquad\square$

Next, we consider the function

$$f_\epsilon(g) = ((\epsilon^2 + |x(g)|^2)^2 + 16|y(g)|^2)^{1/4}, \quad \epsilon \in \mathbb{R}.$$

Lemma 2.4.5. *Let G be a group of Heisenberg type, then for $g \in G$ one has*

$$|Xf_\epsilon(g)|^2 = \frac{|x(g)|^2}{f_\epsilon(g)^2},$$

$$\mathcal{L}f_\epsilon(g) = \frac{Q-1}{f_\epsilon(g)}|Xf_\epsilon(g)|^2 + \frac{m\epsilon^2}{f_\epsilon(g)^3}.$$

Proof. For ease of notation we let $f = f_\epsilon$. Setting $\rho = f^4$ one easily finds

$$|Xf|^2 = \frac{1}{16f^6}|X\rho|^2, \tag{2.61}$$

$$\mathcal{L}f = \frac{1}{4f^3}\left[\mathcal{L}\rho - \frac{3}{4f^4}|X\rho|^2\right]. \tag{2.62}$$

Since

$$X\rho = 2(\epsilon^2 + |x|^2)X(|x|^2) + 16X(|y|^2),$$

using Lemmas 2.4.2 and 2.4.4 we obtain

$$\begin{aligned}
|X\rho|^2 &= 4(\epsilon^2 + |x|^2)^2|X(|x|^2)|^2 + 16^2|X(|y|^2)|^2 \tag{2.63}\\
&\quad + 64(\epsilon^2 + |x|^2) < X(|x|^2), X(|y|^2) > \\
&= 16(\epsilon^2 + |x|^2)^2|x|^2 + 16^2|x|^2|y|^2 \\
&= 16|x|^2\left[(\epsilon^2 + |x|^2)^2 + 16|y|^2\right] = 16|x|^2f^4.
\end{aligned}$$

Substitution in (2.61) gives the first part of the lemma. We compute next $\mathcal{L}\rho$. Applying Lemma 2.4.4 again one finds

$$\mathcal{L}\rho = \mathcal{L}\left((\epsilon^2 + |x|^2)^2\right) + 16\mathcal{L}(|y|^2) = \mathcal{L}\left((\epsilon^2 + |x|^2)^2\right) + 8k|x|^2.$$

On the other hand, Lemma 2.4.2 gives

$$\begin{aligned}
\mathcal{L}\left((\epsilon^2 + |x|^2)^2\right) &= 2|X(|x|^2)|^2 + 2(\epsilon^2 + |x|^2)\mathcal{L}(|x|^2) \\
&= 4(m+2)|x|^2 + 4m\epsilon^2.
\end{aligned}$$

Recalling that the homogeneous dimension of G is $Q = m+2k$, we conclude

$$\mathcal{L}\rho = 4(Q+2)|x|^2 + 4m\epsilon^2. \tag{2.64}$$

Finally, replacing (2.63) and (2.64) in (2.62) we obtain the second part of the lemma. $\qquad\square$

We can now give the proof of Theorem 2.4.1.

Proof of Theorem 2.4.1. With $f = f_\epsilon$ as above and $\epsilon > 0$, we consider the function $w = h(f)$, where $h \in C^2(\mathbb{R})$, and look for conditions on h for which w satisfies the Yamabe type equation

$$\mathcal{L}u = -u^{\frac{Q+2}{Q-2}}. \tag{2.65}$$

Using Lemma 2.4.5 we find

$$\begin{aligned}
\mathcal{L}w &= h''(f)|Xf|^2 + h'(f)\mathcal{L}f \tag{2.66} \\
&= h''(f)\frac{|x|^2}{f^2} + h'(f)\left[\frac{Q-1}{f}|Xf|^2 + \frac{m\epsilon^2}{f^3}\right] \\
&= \left[h''(f) + \frac{Q-1}{f}h'(f)\right]|Xf|^2 + \frac{m\epsilon^2}{f^3}h'(f).
\end{aligned}$$

Formula (2.66) suggests that we choose h such that

$$h''(t) + \frac{Q-1}{t}h'(t) = 0,$$

for each $t \in \mathbb{R}$. The choice $h(t) = \lambda t^{2-Q}, \lambda \in \mathbb{R}$, accomplishes this. Having taken $w = \lambda f^{2-Q}$ we try to satisfy (2.65) for some value of λ. In view of (2.66) this amounts to satisfy the equation

$$\frac{m\epsilon^2}{f^3}h'(f) = -\frac{\lambda^{(Q+2)/(Q-2)}}{f^{(Q+2)}},$$

which reduces to

$$\lambda = \left(m(Q-2)\epsilon^2\right)^{(Q-2)/4}.$$

This completes the proof. $\qquad\square$

Chapter 3

Symmetries of solutions on groups of Iwasawa type

3.1 Intoduction

In this chapter we shall consider groups of Iwasawa type. The goal is to establish symmetry properties of entire solutions of the Yamabe equation

$$\mathcal{L}u = -u^{\frac{Q+2}{Q-2}}, \qquad u \in \overset{o}{\mathcal{D}}{}^{1,2}(\boldsymbol{G}), \quad u \geq 0. \tag{3.1}$$

By Theorem 1.6.9 the above weak solution is actually a smooth bounded function which is everywhere strictly positive, $u > 0$ and $u \in \mathcal{C}^\infty(\Omega)$. The symmetries we are concerned with are obtained by specializing those defined in Definition 1.5.1.

Definition 3.1.1. Let $\boldsymbol{G}$ be a Carnot group of step two with Lie algebra $\mathfrak{g} = V_1 \oplus V_2$. We say that a function $U : \boldsymbol{G} \to \mathbb{R}$ has *partial symmetry* (with respect to a point $g_o \in \boldsymbol{G}$) if there exists a function $u : [0, \infty) \times V_2 \to \mathbb{R}$ such that for every $g = \exp(x(g) + y(g)) \in \boldsymbol{G}$ one has

$$\tau_{g_o} U(g) = u(|x(g)|, y(g)). \tag{3.2}$$

A function U is said to have *cylindrical symmetry* (with respect to $g_o \in \boldsymbol{G}$) if there exists $\phi : [0, \infty) \times [0, \infty) \to \mathbb{R}$ for which

$$\tau_{g_o} U(g) = \phi(|x(g)|, |y(g)|), \tag{3.3}$$

for every $g \in \boldsymbol{G}$.

In [108] the cylindrically symmetric functions were called *biradial*.

The main results of the chapter are Theorem 3.5.1 and Theorem 3.3.1 showing, respectively, that any entire solution with partial symmetry is cylindrically symmetric and the classifing all entire solutions with cylindrical symmetries.

63

The proof of the first result is an adaption of the method of moving hyper-planes due to Alexandrov [2] and Serrin [151]. The moving plane technique was developed further in the two celebrated papers [78], [79] by Gidas, Ni and Nirenberg to obtain symmetry for semi-linear equations with critical growth in $\mathbb{R}^n$ or in a ball. In our proof we incorporate some important simplification of the proof in [79] due to Chen and Li [42]. We mention that a crucial role is played by Theorem 2.4.1 and also by the inversion and the related Kelvin transform introduced by Korányi for the Heisenberg group [114], and subsequently generalized to groups of Heisenberg type in [51], [49].

The proof of the second main result has been strongly influenced by the approach of Jerison and Lee for the Heisenberg group, see Theorem 7.8 in [102]. After a change in the dependent variable, which relates the Yamabe equation to a new non-linear pde in a quadrant of the Poincaré half-plane, one is led to prove that the only positive solutions of the latter are quadratic polynomials of a certain type. Besides Jerison and Lee's paper, the proof has some features of the method of the so-called *P-functions* introduced by Weinberger in [169]. Given a solution u of a certain partial differential equation (pde), such method is based on the construction of a suitable non-linear function of u and *grad u*, a P-function, which is itself solution (or sub-solution) to a related pde, and therefore satisfies a maximum principle. The results in this chapter were proven originally in [76]. In the Theorem 6.1.2 we shall find *all* entire solutions of the Yamabe equation, without assuming a-priori any symmetry, in the case of the seven dimensional quaternionic Heisenberg group.

3.2   The Hopf Lemma

First we recall the strong maximum principles of Bony [19] that will be used. We specialize its statement to the context of Carnot groups, but Theorem 3.2.1 holds in general for Hörmander type operators.

Theorem 3.2.1. *Let Ω be a connected open set in a Carnot group $\mathbf{G}$. Assume that $c \leq 0$ in Ω and that $c \in C(\overline{\Omega})$. If $u \in C^2(\Omega)$ satisfies*

$$\mathcal{L}u + Yu + c\,u \leq 0 \qquad in \quad \Omega,$$

then u cannot achieve a non-positive infimum at an interior point, unless $u \equiv const$ in Ω. Here, Y denotes a smooth vector field on section of $\mathbf{G}$.

The following result constitutes a generalization of the Hopf boundary point lemma, see section 3.2 in [80]. A version for the Heisenberg group first appeared in [26].

Theorem 3.2.2. *In a group of Heisenberg type G let $\Omega \subset G$ be a connected open set possessing an interior gauge ball $B(g_1, R)$ tangent at $g_o \in \partial\Omega$ (by this we mean that $B(g_1, R) \subset \Omega$ and that moreover $g_o \in \partial\Omega \cap \partial B(g_1, R)$). Let $u \in C^2(\Omega)$ be a non-negative solution of*

$$\mathcal{L}u + c\,u \leq 0, \tag{3.4}$$

which is continuous at g_o, and such that

$$u(g_o) = 0, \tag{3.5}$$

$$u(g) > 0, \qquad g \in B(g_1, R) \cap \Omega. \tag{3.6}$$

Assume in addition that $c \in L^\infty(\Omega)$. Let η be any exterior direction at g_o such that $\frac{\partial u}{\partial \eta}(g_o)$ exists, then one has

$$\frac{\partial u}{\partial \eta}(g_o) < 0. \tag{3.7}$$

Proof. We consider $\psi(g) = |x(g)|^2$ and introduce the function $\zeta = e^{-\alpha\psi}u$. A computation based on Lemma 2.4.2 gives

$$\mathcal{L}u = \mathcal{L}(e^{\alpha\psi})\,\zeta + e^{\alpha\psi}\,\mathcal{L}\zeta + 2\,\alpha\,e^{\alpha\psi}\, <X\psi, X\zeta>.$$

Using (3.4) we obtain from the latter equation

$$(4\alpha^2\psi + 2m\alpha + c)\,\zeta + \mathcal{L}\zeta + 2\alpha < X\psi, X\zeta >$$
$$= e^{-\alpha\psi}\,(\mathcal{L}u + cu) \leq 0.$$

This inequality and (3.6) imply in $B(g_o, R) \cap \Omega$

$$\mathcal{L}\zeta + 2\alpha\ <X\psi, X\zeta> \ \leq -[2m\alpha + c]\,\zeta.$$

At this point we choose $\alpha > 0$ such that

$$\alpha \geq \frac{\|c\|_{L^\infty(\Omega)}}{2m},$$

to conclude

$$\mathcal{L}\zeta + 2\alpha\ < X\psi, X\zeta > \ \leq 0 \qquad \text{in}\quad B(g_o, R) \cap \Omega. \tag{3.8}$$

We next use the hypothesis that Ω possesses an interior gauge ball $B(g_1, R)$ tangent at $g_o \in \partial\Omega$. By left-translation we assume without restriction that

$g_1 = e$, where e is the identity in $\boldsymbol{G}$. Recalling (2.12), which we now rewrite as $N(g) = (\psi(g)^2 + 16|y(g)|^2)^{1/4}$, we introduce the auxiliary function

$$h(g) = e^{-MN^2(g)} - e^{-MR^2}$$

on the ring $A = A(R,r) = B(e,R)\backslash\overline{B}(e,r)$, where $0 < r < R$ has been fixed. The constant $M > 0$ will be chosen shortly. An elementary computation, using the fact that $\mathcal{L}(N^{2-Q})(g) = 0$ for every $g \neq e$, see (2.13), gives

$$\mathcal{L}(N^2)(g) = \frac{Q}{2N^2(g)} \, |X(N^2)(g)|^2, \qquad \text{for} \quad g \neq e. \tag{3.9}$$

Formula (3.9) allows to find

$$\mathcal{L}h + 2\alpha \; < X\psi, Xh > \tag{3.10}$$

$$= M \, e^{-MN^2} \left[(M - \frac{Q}{2N^2}) \, |X(N^2)|^2 - 2\alpha \; < X\psi, X(N^2) > \right].$$

Using Theorem 2.4.4 we find

$$< X\psi, X(N^2) > = \frac{1}{2} \, N^{-2} \, [2\psi|X\psi|^2 + < X\psi, X(|y|^2) >]$$

$$= 4N^{-2} \, \psi^2 = 4N^2 \, |XN|^4, \tag{3.11}$$

since in a group of Heisenberg type one has $|XN|^2 = N^{-2}\psi$, which follows easily from Lemmas 2.4.2 and 2.4.4. The identity (3.11) allows to conclude that choosing $M > 0$ sufficiently large in (3.10) one obtains

$$\mathcal{L}h + 2\alpha \; < X\psi, Xh > \geq 0 \qquad \text{in} \quad A. \tag{3.12}$$

The continuity of u in Ω and the compactness of $\partial B(e,r)$ implies the existence of $\epsilon > 0$ such that the function $\zeta - \epsilon h \geq 0$ on $\partial B(e,r)$. This inequality continues to hold on $\partial B(e,R)$ since $h = 0$ on that set. By (3.8), (3.12) and Theorem 3.2.1 we conclude $\zeta - \epsilon h \geq 0$ in A. Since u, ζ and h vanish in g_o we conclude

$$\frac{\partial u}{\partial \eta}(g_o) = e^{\alpha\psi(g_o)} \, \frac{\partial\zeta}{\partial\eta}(g_o) \leq \epsilon \, e^{\alpha\psi(g_o)} \, \frac{\partial h}{\partial\eta}(g_o),$$

where η is any direction such that $< \eta, \mathbf{N} > (g_o) > 0$, with $\mathbf{N}$ being the exterior unit normal to $\partial B(e,R)$. At this point the conclusion follows by observing that the function $N(g)$ is homogeneous of degree one and therefore denoting by Z the infinitesimal generator of group dilations we have $ZN(g) = N(g)$ for every $g \neq e$. This identity implies in particular that the Riemannian gradient of $N(g)$, ∇N, never vanishes in $\boldsymbol{G} \setminus \{e\}$. Since $\partial B(e,R)$ is a level set of N and ∇N is directed outward, we infer

$$\frac{\partial h}{\partial\eta}(g_o) = -2MRe^{-MR^2} \, \frac{\partial N}{\partial\eta}(g_o) < 0.$$

This completes the proof of the theorem. $\qquad\qquad\qquad\square$

Corollary 3.2.3. *Let $u \in C^2(\Omega)$ be a non-negative solution of* (3.4) *in $\Omega \subset G$, where G is a group of Heisenberg type, then u cannot become equal to zero at an interior point without being identically zero in Ω.*

Proof. The proof follows the lines of its elliptic counterpart, see [[80], Theorem 3.5]. Assume by contradiction that u vanishes at a point inside Ω without being identically zero. Define $\Omega^+ \overset{def}{=} \{g \in \Omega \mid u(g) > 0\}$, which is non-empty according to the assumption, and satisfies $\Omega^+ \subset \Omega$, $\partial\Omega^+ \cap \Omega \neq \varnothing$. Let $g_o \in \Omega^+$ be closer to $\partial\Omega^+$ than to $\partial\Omega$, with respect to the gauge distance. Consider the largest gauge ball $B \subset \Omega^+$ having g_o as its center. Then $u(g) = 0$ for some point $g \in \partial B$, while $u > 0$ in B. By left-translation we can assume that $g = e$, the group identity. Since g is a point of an interior minimum on Ω, the Riemannian gradient at g must vanish. This is a contradiction with Theorem 3.2.2, by considering for example the derivative along the generator Z of the group dilations. $\quad\square$

We end this section with a simple geometric result which is used in the application of the method of moving hyper-planes, see Lemma 2.2 in [91].

Proposition 3.2.4. *If a connected compact surface in $\mathbb{R}^k$ has the property that for every direction $\xi \in \mathbb{R}^k$ there exists a hyper-plane Π_ξ perpendicular to ξ, such that S is symmetric with respect to Π_ξ, then S is a Euclidean sphere.*

Proof. In the proof distance refers to the Euclidean distance in $\mathbb{R}^k$. Let A_1, $A_2 \in S$ be two points at a distance equal to the (Euclidean) diameter of S. Let O be the middle point of the segment $A_1 A_2$ and π be the hyperplane through O that is at an equal distance from the points A_1 and A_2. We note that π is the plane of symmetry in the direction perpendicular to $A_1 A_2$. If not, by considering also the reflexions of A_1 and A_2 with respect to π, we can obtain points on S at a distance greater than the diameter of S, which is a contradiction. Next, we consider an arbitrary plane of symmetry, π', and show that it passes through the point O as well. Let α_1 and α_2 be the hyperplanes, correspondingly through A_1 and A_2, which are parallel to π. From the choice of A_1 and A_2, the surface S lies between α_1 and α_2. If we suppose $O \notin \pi \cap \pi'$, then at least one of the reflexion points of A_1 and A_2 with respect to π' will not belong to the slab between α_1 and α_2. This is a contradiction, since S is invariant under reflexions with respect to π'. Thus, S coincides with the sphere through A_1 and A_2, and center O. The proof is finished. $\quad\square$

3.3   The partially symmetric solutions have cylindrical symmetry

In this section we use the method of moving hyper-planes to prove the following Theorem.

Theorem 3.3.1. *Let G be an Iwasawa group. Suppose $U \not\equiv 0$ is an entire solution of the Yamabe equation (3.1). If U has partial symmetry, then U has cylindrical symmetry.*

We will use the letters α, α' to index coordinates in the first layer, and β, β' for indexing the coordinates in the center, so that we have $1 \leq \alpha$, $\alpha' \leq m$ and $1 \leq \beta$, $\beta' \leq k$. Unless explicitly said otherwise, we shall use the same letter for a function f defined on G and for the corresponding function $f \circ \exp$ defined on $\mathfrak{g} \approx \mathbb{R}^m \times \mathbb{R}^k$. In the next proposition we express the sub-Laplacian in the exponential coordinates, see also [55].

Proposition 3.3.2. *For every $\beta = 1, ..., k$, let T_β denote the vector field*

$$T_\beta = \sum_{\alpha,\alpha'=1}^{m} x_{\alpha'} < [X_{\alpha'}, X_\alpha], Y_\beta > \frac{\partial}{\partial x_\alpha}.$$

Using the exponential coordinates we have the following formula for the sub-Laplacian of a function $u : G \to \mathbb{R}$

$$\mathcal{L}u(g) = \triangle_x u(g) + \sum_{\beta=1}^{k} T_\beta \frac{\partial u}{\partial y_\beta}(g) + \frac{1}{4}|x(g)|^2 \, \triangle_y u(g). \qquad (3.13)$$

In (3.13) we have respectively denoted with $\triangle_x$ and $\triangle_y$ the standard Laplacian in $\mathbb{R}^m$ and $\mathbb{R}^k$.

Proof. To avoid confusion, in the course of the proof we will keep the distinct notation $v(x, y)$ for the function $u \circ \exp$ on the Lie algebra. Here, we note explicitly that $g = \exp(\xi(g)) = \exp(\xi_1(g) + \xi_2(g)) = \exp\left(\sum_{\alpha=1}^{m} x_\alpha(g)X_\alpha + \sum_{\beta=1}^{k} y_\beta(g)Y_\beta\right)$. By the Baker-Campbell-Hausdorff formula we have for every $\alpha = 1, ..., m$

$$X_\alpha u(g) = \frac{\partial u}{\partial t}(\exp(\xi + tX_\alpha + \frac{t}{2}[\xi, X_\alpha]))\,|_{t=0} = \frac{\partial v}{\partial t}(\xi + tX_\alpha + \frac{t}{2}[\xi, X_\alpha])\,|_{t=0}$$

$$= \sum_{\alpha'} < X_\alpha + \frac{1}{2}[\xi, X_\alpha], X_{\alpha'} > \frac{\partial v}{\partial x_{\alpha'}} + \sum_{\beta} < X_\alpha + \frac{1}{2}[\xi, X_\alpha], Y_\beta > \frac{\partial v}{\partial y_\beta}$$

$$= \frac{\partial v}{\partial x_\alpha} + \frac{1}{2}\sum_{\beta} < [\xi, X_\alpha], Y_\beta > \frac{\partial v}{\partial y_\beta}, \qquad (3.14)$$

where we have used the orthonormality of the involved vectors and the stratification of $\mathfrak{g}$. We also note that
$$< [\xi, X_\alpha], Y_\beta > \; = \; < [\xi_1, X_\alpha], Y_\beta > .$$
We thus obtain from (3.14)
$$X_\alpha u = \frac{\partial v}{\partial x_\alpha} + \frac{1}{2} \sum_\beta \; < [\xi_1, X_\alpha], Y_\beta > \; \frac{\partial v}{\partial y_\beta}. \tag{3.15}$$
Notice that $[X_\alpha, X_\alpha] = 0$ gives
$$\frac{\partial}{\partial x_\alpha} < [\xi_1, X_\alpha], Y_\beta > \; = \sum_{\alpha'} \frac{\partial}{\partial x_\alpha} x_{\alpha'} \; < [X_{\alpha'}, X_\alpha], Y_\beta > \; = 0. \tag{3.16}$$
Obviously, we have also
$$\frac{\partial}{\partial y_{\beta'}} < [\xi_1, X_\alpha], Y_\beta > \; = 0. \tag{3.17}$$
Applying (3.15) twice and using (3.16) and (3.17) we find
$$X_\alpha^2 u(g) = \Big(\frac{\partial}{\partial x_\alpha} + \frac{1}{2} \sum_{\beta'} \; < [\xi_1, X_\alpha], Y_{\beta'} > \; \frac{\partial}{\partial y_{\beta'}}\Big)$$
$$\Big(\frac{\partial v}{\partial x_\alpha} + \frac{1}{2} \sum_\beta \; < [\xi_1, X_\alpha], Y_\beta > \; \frac{\partial v}{\partial y_\beta}\Big)$$
$$= \frac{\partial^2 v}{\partial x_\alpha^2} + \sum_\beta \; < [\xi_1, X_\alpha], Y_\beta > \; \frac{\partial^2 v}{\partial x_\alpha \partial y_\beta}$$
$$+ \frac{1}{4} \sum_{\beta, \beta'} \; < [\xi_1, X_\alpha], Y_\beta > < [\xi_1, X_\alpha], Y_{\beta'} > \; \frac{\partial^2 v}{\partial y_\beta \partial y_{\beta'}}.$$
Summing in α we obtain
$$\mathcal{L}u = \triangle_x v + \sum_\beta T_\beta \frac{\partial v}{\partial y_\beta} + \frac{1}{4} \sum_{\alpha, \beta, \beta'} \; < [\xi_1, X_\alpha], Y_\beta > < [\xi_1, X_\alpha], Y_{\beta'} > \; \frac{\partial^2 v}{\partial y_\beta \partial y_{\beta'}}.$$
Now we use (2.5) and orthonormality to further reduce the last term in the right hand-side of the latter expression
$$\sum_{\alpha, \beta, \beta'} \; < [\xi_1, X_\alpha], Y_\beta > \; < [\xi_1, X_\alpha], Y_{\beta'} > \; \frac{\partial^2 v}{\partial y_\beta \partial y_{\beta'}}$$
$$= \sum_{\beta \beta'} \Big(\sum_\alpha \; < J(Y_\beta)\xi_1, X_\alpha > \; < J(Y_{\beta'})\xi_1, X_\alpha >\Big) \frac{\partial^2 v}{\partial y_\beta \partial y_{\beta'}}$$
$$= \sum_{\beta, \beta'} \; < J(Y_\beta)\xi_1, J(Y_{\beta'})\xi_1 > \; \frac{\partial^2 v}{\partial y_\beta \partial y_{\beta'}}$$
$$= \sum_{\beta, \beta'} \; < Y_\beta, Y_{\beta'} > |x|^2 \frac{\partial^2 v}{\partial y_\beta \partial y_{\beta'}} = |x|^2 \triangle_y v.$$
This completes the proof. $\qquad\qquad\qquad\qquad\qquad\qquad\qquad\qquad\square$

The next two results are direct consequences of Proposition 3.3.2.

Proposition 3.3.3. *Suppose that U has the form $U(g) = u(|x(g)|, y(g))$. For every $\beta = 1, ..., k$ one has $T_\beta u \equiv 0$ and therefore (3.13) gives*

$$\mathcal{L}U(g) = \triangle_x u(g) + \frac{|x(g)|^2}{4} \triangle_y u(g). \tag{3.18}$$

In particular, the vector fields T_β, $\beta = 1, ..., k$, are tangential to domains with cylindrical symmetry.

Proof. One has

$$
\begin{aligned}
T_\beta u &= \Big(\sum_{\alpha,\alpha'} x_{\alpha'} < [X_{\alpha'}, X_\alpha], Y_\beta > \frac{\partial}{\partial x_\alpha}\Big) u \\
&= \sum_{\alpha,\alpha'} \frac{x_{\alpha'} x_\alpha}{r} < J(Y_\beta) X_{\alpha'}, X_\alpha > \frac{\partial u}{\partial r} = \frac{1}{r} < J(Y_\beta)\xi_1, \xi_1 > \frac{\partial u}{\partial r} = 0,
\end{aligned}
$$

where the last equality is justified by (2.4). The proof is completed.   □

Proposition 3.3.4. *Suppose that U has the form $U(g) = u(|x(g)|, y(g))$. One has the following formula for the horizontal gradient of U*

$$|XU(g)|^2 = |\nabla_x u(g)|^2 + \frac{|x(g)|^2}{4} |\nabla_y u(g)|^2. \tag{3.19}$$

Proof. Taking the squares and summing in α equation (3.15) we obtain

$$|Xu|^2 = |\nabla_x u|^2 + \frac{1}{4} \sum_\alpha \Big(\sum_\beta < [\xi_1, X_\alpha], Y_\beta > \frac{\partial u}{\partial y_\beta}\Big)^2 + \sum_\beta T_\beta u \frac{\partial u}{\partial y_\beta}.$$

Since u has partial symmetry $T_\beta u = 0$ and the last term in the above equality is zero. The second term is computed by using the orthogonality of the map J and the orthonormality of the vector fields X_α and Y_β,

$$
\begin{aligned}
\sum_\alpha \Big(\sum_\beta < [\xi_1, X_\alpha], Y_\beta > \frac{\partial u}{\partial y_\beta}\Big)^2 &= \sum_\alpha < [\xi_1, X_\alpha], \sum_\beta \frac{\partial u}{\partial y_\beta} Y_\beta >^2 \\
&= \sum_\alpha < J(\sum_\beta \frac{\partial u}{\partial y_\beta} Y_\beta)\xi_1, X_\alpha >^2 = |J(\sum_\beta \frac{\partial u}{\partial y_\beta} Y_\beta)\xi_1|^2 \\
&= |x(g)|^2 |\sum_\beta \frac{\partial u}{\partial y_\beta} Y_\beta|^2 = |x(g)|^2 |\nabla_y u|^2.
\end{aligned}
$$

The proof is complete.   □

Remark 3.3.5. Proposition 3.3.3 underlines the important connection between the sub-Laplacian on a group of Heisenberg type and the Baouendi-Grushin operator

$$\triangle_x + \frac{|x(g)|^2}{4}\,\triangle_y$$

acting on functions which possess partial symmetry.

After the above preliminaries our next goal is to prove Theorem 3.3.1. Before starting the proof however we introduce the relevant notation and develop some preparatory results. In the remainder of this section we will always identify a point $g = \exp(\xi_1 + \xi_2) \in G$ with its exponential coordinates $(x, y) = (x(g), (y(g)) \in \mathbb{R}^m \times \mathbb{R}^k$, where $x = x(g) = (x_1(g), ..., x_m(g))$ and $y = y(g) = (y_1(g), ..., y_k(g))$ are defined by (2.49) and (2.50). For any $\lambda \in \mathbb{R}$ we consider the *characteristic half-spaces* in G

$$\Sigma_\lambda = \{g = (x, y) \in G \mid y_1 < \lambda\}, \qquad \lambda < 0, \qquad (3.20)$$

and

$$\Sigma_\lambda = \{g = (x, y) \in G \mid y_1 > \lambda\}, \qquad \lambda > 0. \qquad (3.21)$$

We denote by T_λ the *characteristic hyper-planes* $\partial\Sigma_\lambda = \{g = (x, y) \in G \mid y_1 = \lambda\}$. For any $g \in \Sigma_\lambda$ we let g^λ be the *symmetric* point with respect to the hyperplane T_λ, i.e., $g^\lambda = (x(g), 2\lambda - y_1(g), y_2(g), \ldots, y_k(g))$. Finally, we let $g_\lambda = (0, 2\lambda, 0, \ldots, 0) \in \Sigma_\lambda$ be the reflexion of $(0, 0) \in \mathbb{R}^m \times \mathbb{R}^k$ with respect to T_λ.

Next, we assume that $u \not\equiv 0$ is an entire solution to the problem (3.1). From Theorem 1.6.9 we know that $u \in L^\infty(G) \cup C^\infty(G)$. From the strong maximum principle, Theorem 3.2.1, we have also $u > 0$ on G. Consider in G^* the Kelvin transform of u, $v(g) = N(g)^{2-Q}u(\sigma(g))$, as in Definition 2.3.2. Clearly,

$$\lim_{N(g)\to\infty} N(g)^{Q-2}\,v(g) = u(e) > 0. \qquad (3.22)$$

Setting $v^\lambda(g) = v(g^\lambda)$, we define

$$\bar{w}_\lambda(g) = \frac{v^\lambda(g) - v(g)}{K_\epsilon(g)} \stackrel{def}{=} \frac{w_\lambda(g)}{K_\epsilon(g)}, \qquad g \in \bar\Sigma_\lambda, \qquad (3.23)$$

where $K_\epsilon(g)$ is the function in Theorem 2.4.1. We observe that $\bar{w}_\lambda \equiv 0$ on T_λ. It is clear that v^λ is singular in $g = g_\lambda \in \Sigma_\lambda$ and that v is singular in $g = e$. However, thanks to Theorem 2.3.5, Lemma 2.3.6 and Theorem 2.3.7 we can remove the singularities so that v^λ and v become entire solutions to (3.1). This guarantees that $\bar{w}_\lambda$ is now globally defined on $\boldsymbol{G}$. We note that for every fixed λ one has

$$\lim_{N(g)\to\infty} \bar{w}_\lambda(g) = 0. \tag{3.24}$$

To prove (3.24) we first observe that

$$N(g^\lambda)^{2-Q} - N(g)^{2-Q} = N(g)^{2-Q}\, \Omega_\lambda(g), \tag{3.25}$$

where $|\Omega_\lambda(g)| \to 0$ as $N(g) \to \infty$. From (3.25) one easily infers that

$$\lim_{N(g)\to\infty} N(g)^{Q-2}\, v^\lambda(g) = u(e) > 0. \tag{3.26}$$

We now write

$$v^\lambda(g) - v(g) = N(g^\lambda)^{2-Q}\, [u(\sigma(g^\lambda)) - u(\sigma(g))] + [N(g^\lambda)^{2-Q} - N(g)^{2-Q}]\, u(\sigma(g)). \tag{3.27}$$

Using (3.25) in (3.27) we obtain (3.24).

To apply the method of moving hyper-planes we establish next a result analogous to Lemma 2.1 in [42].

Lemma 3.3.6.

    (i) *If* $\inf_{\Sigma_\lambda} \bar{w}_\lambda < 0$, *then the infimum is achieved.*

    (ii) *There exists* $R_o > 0$ *independent of* λ *such that, if* $g_o \in \Sigma_\lambda$ *is a point at which a strictly negative* $\inf_{\Sigma_\lambda} \bar{w}_\lambda$ *is attained, then* $N(g_o) < R_o$. *Furthermore, for all* $|\lambda| \geq R_0^2$ *we have* $\bar{w}_\lambda \geq 0$ *on* Σ_λ.

Proof. The proof of (i) is easy. Suppose that for a certain λ one has

$$\inf_{\Sigma_\lambda} \bar{w}_\lambda = m_\lambda < 0.$$

Consider the set $A_\lambda = \{g \in \bar{\Sigma}_\lambda \mid \bar{w}_\lambda(g) \leq m_\lambda/2\}$. Equation (3.24) and $w_\lambda \equiv 0$ on T_λ imply that A_λ is a compact set. By the continuity of $\bar{w}_\lambda$ on $\boldsymbol{G}$ we conclude the validity of (i).

To prove (ii) we begin by observing that thanks to Proposition 3.3.3 we have

$$\mathcal{L}v^\lambda(g) = \mathcal{L}v(g^\lambda). \tag{3.28}$$

Using (3.28) and the mean value theorem we find in Σ_λ

$$\mathcal{L}w_\lambda(g) = \mathcal{L}v(g^\lambda) - \mathcal{L}v(g) = v^{2^*-1}(g) - v^{2^*-1}(g^\lambda) = -c_\lambda(g)\, w_\lambda(g), \tag{3.29}$$

where

$$c_\lambda(g) = (2^* - 1)\, \psi_\lambda^{2^*-2}(g), \tag{3.30}$$

with $\psi_\lambda(g)$ a real number between $v(g^\lambda)$ and $v(g)$. The equation satisfied by $\bar{w}_\lambda$ can be obtained from (3.29) and from $w_\lambda = K_\epsilon \bar{w}_\lambda$,

$$\mathcal{L}\bar{w}_\lambda + \frac{2}{K_\epsilon} < XK_\epsilon, X\bar{w}_\lambda > + \left(c_\lambda + \frac{\mathcal{L}K_\epsilon}{K_\epsilon}\right)\bar{w}_\lambda = 0. \tag{3.31}$$

From (3.22) we infer the existence of $C_o > 0$ such that

$$v(g) \leq \frac{C_o}{(1 + N(g))^{Q-2}}, \qquad g \in G. \tag{3.32}$$

We now show that there exist $\epsilon = \epsilon(C_o) > 0$ and $R_o = R_o(C_o) > 0$ such that

$$c_\lambda(g) + \frac{\mathcal{L}K_\epsilon}{K_\epsilon}(g) < 0, \tag{3.33}$$

for $N(g) \geq R_o$ whenever $\bar{w}_\lambda(g) < 0$. For λ such that

$$\Omega_\lambda \overset{def}{=} \{g \in \Sigma_\lambda \mid \bar{w}_\lambda(g) < 0\} \neq \varnothing$$

consider (3.31) on Ω_λ. If $g \in \Omega_\lambda$, we have $v(g^\lambda) < v(g)$ and thus (3.32) gives

$$v(g^\lambda) < v(g) \leq \frac{C_o}{N(g)^{Q-2}}. \tag{3.34}$$

Since ψ_λ is between $v(g)$ and $v(g^\lambda)$ we conclude from (3.30) and (3.34) that

$$c_\lambda(g) \leq \frac{(2^* - 1)C_o^{2^*-2}}{N(g)^{(Q-2)(2^*-2)}} = \frac{C_1}{N(g)^4}. \tag{3.35}$$

Thanks to Theorem 2.4.1 we have

$$\frac{\mathcal{L}K_\epsilon(g)}{K_\epsilon(g)} = -K_\epsilon(g)^{2^*-2} = -\frac{m(Q-2)\epsilon^2}{(\epsilon^2 + |x(g)|^2)^2 + 16|y(g)|^2}. \tag{3.36}$$

From (3.35) and (3.36) we see that

$$\begin{aligned}
c_\lambda(g) + \frac{\mathcal{L}K_\epsilon(g)}{K_\epsilon(g)} &\leq \frac{C_1}{N(g)^4} - \frac{m(Q-2)\epsilon^2}{(\epsilon^2 + |x(g)|^2)^2 + 16|y(g)|^2} \\
&= \frac{C_1 N(g)^4 + C_1(\epsilon^4 + 2\epsilon^2|x(g)|^2) - m(Q-2)\epsilon^2 N(g)^4}{N(g)^4\big((\epsilon^2 + |x(g)|^2)^2 + 16|y(g)|^2\big)} \\
&\leq \frac{(C_1 - m(Q-2)\epsilon^2)N(g)^4 + 2C_1\epsilon^2 N(g)^2 + C_1\epsilon^4}{N(g)^4\big((\epsilon^2 + |x(g)|^2)^2 + 16|y(g)|^2\big)}.
\end{aligned}$$

If we choose $\epsilon^2 = \frac{2C_1}{m(Q-2)}$ in the latter inequality, the coefficient of $N(g)^4$ is negative, and it is then clear that we can fulfill (3.33) for $N(g) \geq R_o$ for some $R_o = R_o(C_1) > 0$. From Theorem 3.2.1 we conclude that $\bar{w}_\lambda$ cannot achieve a negative infimum for $N(g) \geq R_o$ in Ω_λ. This proves the first part of (ii). At this point we observe that

$$N(g) \geq \sqrt{|\lambda|}, \qquad \text{for every} \quad g \in \Sigma_\lambda.$$

It is therefore clear from the above argument that if we take $|\lambda| \geq R_o^2$, then $\bar{w}_\lambda$ cannot achieve a negative infimum in Σ_λ. This completes the proof of Lemma 3.3.6.   □

We are now ready to present the proof of Theorem 3.3.1.

Proof of Theorem 3.3.1. Let $u(x,y) \overset{def}{=} \tau_{g_0} U(x(g),y(g)) = u(|x(g)|, y(g))$ and v be the Kelvin transform of u. Since $\mathcal{L}$ is a translation invariant operator, u is an entire solution to (3.1). For ease of notation we are using the same letter to denote functions on G, which have partial symmetry with respect to the identity element of G, and the corresponding symmetric part defined on $[0,\infty) \times \mathbb{R}^k$, see Definition 3.1.1. As already mentioned in the paragraph after (3.23), v is also an entire solution on G. Furthermore, from (2.33) it is easy to see that the Kelvin transform of a function that has partial symmetry with respect to the identity element of G is a function with partial symmetry with respect to the identity element as well. The first step of the proof is to show that v has cylindrical symmetry.

Let $\lambda_o = \sup\{\lambda \leq 0 \mid \bar{w}_\lambda \geq 0 \text{ in } \Sigma_\lambda\}$. Clearly $\lambda_o \leq 0$. Assume first that $\lambda_o < 0$. We want to show that $\bar{w}_{\lambda_o} \equiv 0$. Suppose the contrary holds. Since

$$\mathcal{L}w_{\lambda_o} + c_{\lambda_o} w_{\lambda_o} = 0 \qquad \text{in} \quad \Sigma_{\lambda_o}, \tag{3.37}$$

with c_{λ_o} bounded and $w_{\lambda_o} \geq 0$, Theorem 3.2.2 implies that either $w_{\lambda_o} > 0$, or $w_{\lambda_o} \equiv 0$. Since we are assuming $w_{\lambda_o} \not\equiv 0$, we conclude $w_{\lambda_o} > 0$. This implies $\bar{w}_{\lambda_o} > 0$ in Σ_{λ_o}. The maximality of λ_o allows to find a sequence $\lambda_k \searrow \lambda_o$ and points $g_k \in \Sigma_{\lambda_k}$ such that

$$\bar{w}_{\lambda_k}(g_k) < 0. \tag{3.38}$$

Without restriction we can suppose $\bar{w}_{\lambda_k}(g_k) = \inf_{\Sigma_{\lambda_k}} \bar{w}_{\lambda_k}$, since by (i) of Lemma 3.3.6 the infimum is achieved when it is strictly negative. We thus have

$$d\bar{w}_{\lambda_k}(g_k) = 0. \tag{3.39}$$

In the proof of Lemma 3.3.6 we saw that the sequence $\{g_k\}$ is uniformly bounded, in fact $N(g_k) \leq R_o$. Possibly passing to a subsequence, we can thus assume that $g_k \to g_o \in \bar{\Sigma}_{\lambda_o}$. By continuity from (3.38) and (3.39) we have $\bar{w}_{\lambda_o}(g_o) \leq 0$ and $d\bar{w}_{\lambda_o}(g_o) = 0$. Since $\bar{w}_{\lambda_o} > 0$ in Σ_{λ_o}, it must be $g_o \in T_{\lambda_o}$. Finally, $\bar{w}_{\lambda_o} > 0$ and $d\bar{w}_{\lambda_o}(g_o) = 0$ contradict Theorem 3.2.2 by considering the derivative along any direction non-tangential to the boundary. This shows that when $\lambda_o < 0$ we have $w_{\lambda_o} \equiv 0$, i.e., v is symmetric with respect to the hyperplane T_{λ_o}.

If $\lambda_o = 0$ we can repeat the above reasoning starting from $\lambda = +\infty$ and then either stop at some $\lambda_1 > 0$, or at $\lambda_1 = 0$. In the former case we can finish as above. In the latter we combine the conclusions of the two cases to see that $v(g) > v(g^\lambda)$ and $v(g) < v(g^\lambda)$, i.e., $v(g) = v(g^\lambda)$ for any g and g^λ symmetric with respect to the hyperplane $y_1 = 0$. In either case, we conclude that v is symmetric with respect to T_λ for some λ. We note also that the restriction of v to *lines perpendicular* to T_λ is a monotonically decreasing function of the Euclidean distance to T_λ. In order to see this, suppose $\lambda_o < 0$ so that $T_\lambda = T_{\lambda_o}$. Consider an arbitrary line l perpendicular to T_{λ_o} and let $P_1, P_2 \in \Sigma_{\lambda_o} \cap l$, with P_2 between P_1 and the intersection of T_{λ_o} and l. By considering the plane $T_{\bar{\lambda}}$ with respect to which P_1 and P_2 are symmetric, using also the definition of λ_o, we see that $v(P_1) < v(P_2)$. Arguing similarly in the case of $\lambda_o \geq 0$ we see that v has the described monotonicity, when restricted to any line perpendicular to T_{λ_o}.

From Proposition 3.3.3, $\mathcal{L}$ is an operator invariant with respect to rotations in the center, when restricted to partially symmetric functions. Since v has partial symmetry, v is invariant under such rotations. The previous arguments show that for every direction in the center, $\mathbb{R}^k$, there exists a hyperplane $T = T_{\lambda_o} \cap \mathbb{R}^k$ perpendicular to it, such that for every $r > 0$, $v(r, \cdot)$ is symmetric with respect to T. We note explicitly that T is independent of r. In addition, v has the above monotonicity on lines perpendicular to T. Since v is a continuous function and $v(g) \to 0$ when $N(g) \to \infty$, every level set is compact. Therefore, using also the monotonicity of v, for every $r \geq 0$ and every regular value a, the level set $v(r, \cdot) = a$ is a connected closed hypersurface of $\mathbb{R}^k$, when it is non-empty. Furthermore, from the symmetry of v, every level set of the function $v(r, \cdot)$ defined on $\mathbb{R}^k$, is symmetric with respect to the hyperplane T. In view of Proposition 3.2.4 we infer that every level set is a sphere. Spheres corresponding to different regular values a are concentric, for otherwise we can argue as follows. Let $O_1 \neq O_2$ be the centers of two such non-concentric spheres. Let us consider the plane of symmetry, which is perpendicular to the direction of $O_1 O_2$. Using again the monotonicity of v, we have on one hand that it should pass through O_1, while on the other it should pass through O_2, which is impossible. Finally, for any $b > 0$ consider the level set $\Lambda_b \overset{def}{=} \{v > b\}$. Clearly $\Lambda_b = \underset{a > b}{\cup} \Lambda_a$ and from Sard's theorem there exists a sequence $\{a_k\}$ of regular values such that $a_k \searrow b$. Since the level sets corresponding to regular values are Euclidean balls in $\mathbb{R}^k$, their union is a ball as well. This shows that $v(r, \cdot)$ is a radial function of its argument, after choosing suitably the origin of $\mathbb{R}^k$. Since the planes of symmetry are independent of r, the above choice of the origin of $\mathbb{R}^k$ is independent of r as well. In other words v is a cylindrical function.

The final step is to reverse the roles of u and v, using the properties of the Kelvin transform. In the beginning of the proof we noted that v is an entire solution that has partial symmetry with respect to the identity element. Since the Kelvin transform is an involution, from the first step of the proof we see that u has cylindrical symmetry, i.e., there exists an $h_o \in G$ (in fact h_o belongs to the center of G) such that $\tau_{h_o} u = \phi(|x(g)|, |y(g)|)$. Therefore,

$$\tau_{h_o g_o} U = \tau_{h_o} \tau_{g_o} U = \phi(|x(g)|, |y(g)|)$$

and the proof of Theorem 3.3.1 is complete. $\qquad\qquad\qquad\qquad\square$

3.4 Determination of the cylindrically symmetric solutions of the Yamabe equation

In this section we establish the uniqueness, modulo group translations and dilations, of the positive solutions with cylindrical symmetry to the Yamabe equation (3.1) Our main objective is the following Theorem.

Theorem 3.4.1. *Let $U \not\equiv 0$ be an entire solution to (3.1) in a group of Iwasawa type G and suppose that U has cylindrical symmetry. There exists $\epsilon > 0$ such that up to a left-translation (1.28) one has*

$$U(g) = \left(\frac{m(Q-2)\epsilon^2}{(\epsilon^2 + |x(g)|^2)^2 + 16|y(g)|^2} \right)^{\frac{Q-2}{4}}.$$

We begin with some preliminary reductions. The first observation is that if we let $v = \lambda u$, then by choosing

$$\lambda = \left(\frac{Q-2}{4}\right)^{-(Q-2)/2}, \tag{3.40}$$

we are reduced to consider the equation

$$\mathcal{L}v = -\left(\frac{Q-2}{4}\right)^2 v^{(Q+2)/(Q-2)}. \tag{3.41}$$

Next, we introduce the function $\Phi = v^{-4/(Q-2)} = h(v)$. Since

$$\mathcal{L}\Phi = h''(v)|Xv|^2 + h'(v)\mathcal{L}v,$$

we easily find that Φ must satisfy the equation

$$\mathcal{L}\Phi = \left(\frac{Q-2}{4} + 1\right) \frac{|X\Phi|^2}{\Phi} + \frac{Q-2}{4}. \tag{3.42}$$

At this point we assume that u, and therefore Φ, have cylindrical symmetry with respect to the identity, i.e., there exists a function $\phi : [0, \infty) \times [0, \infty) \to \mathbb{R}^+$ such that we can write with $g = \exp(\xi_1 + \xi_2) \in \boldsymbol{G}$

$$\Phi(g) = \phi(|\xi_1|, |\xi_2|). \tag{3.43}$$

By Proposition 3.3.3 we see that the equation (3.42) now becomes

$$\Delta_{\xi_1} \phi + \frac{|\xi_1|^2}{4} \Delta_{\xi_2} \phi = \left(\frac{Q-2}{4} + 1\right) \frac{1}{\phi} \left(|\nabla_{\xi_1}\phi|^2 + \frac{|\xi_1|^2}{4} |\nabla_{\xi_2}\phi|^2\right) + \frac{Q-2}{4}. \tag{3.44}$$

Passing to the spherical coordinates $r = |\xi_1|, s = |\xi_2|$, we obtain from (3.44)

$$\phi_{rr} + \frac{m-1}{r} \phi_r + \frac{r^2}{4} \left(\phi_{ss} + \frac{k-1}{s} \phi_s\right) \tag{3.45}$$
$$= \left(\frac{Q-2}{4} + 1\right) \frac{1}{\phi} \left(\phi_r^2 + \frac{r^2}{4} \phi_s^2\right) + \frac{Q-2}{4}.$$

We now let

$$y = \frac{r^2}{4}, \qquad x = s, \tag{3.46}$$

obtaining from (3.45)

$$\phi_{xx} + \phi_{yy} + \frac{m}{2y} \phi_y + \frac{k-1}{s} \phi_s \tag{3.47}$$
$$= \left(\frac{Q-2}{4} + 1\right) \frac{\phi_x^2 + \phi_y^2}{\phi} + \frac{Q-2}{4} \frac{1}{y}.$$

Defining the integers

$$a = k - 1 \geq 0, \qquad b = \frac{m}{2} \geq 1, \qquad n = a + b \geq 1, \qquad (3.48)$$

and recalling that $Q = m + 2k$, we finally re-write equation (3.47) as follows

$$\Delta \phi = \frac{n+2}{2} \frac{|\nabla \phi|^2}{\phi} - \frac{a}{x} \phi_x - \frac{b}{y} \phi_y + \frac{n}{2y}, \qquad (3.49)$$

in $\Omega = \{(x, y) \in \mathbb{R}^2 \mid x > 0, y > 0\}$. We remark explicitly at this point that, without loss of generality, we can assume that $k \geq 2$, and therefore $a \geq 1$. In fact, the case $k = 1$ corresponds to the Heisenberg group $\mathbb{H}^n$, and it has already been treated by Jerison and Lee in [102]. We now introduce the quantities

$$F = f - f^*, \qquad G = g + g^*, \qquad (3.50)$$

where

$$f = 2 < \nabla \phi, \nabla \phi_x > -2\delta \, \phi_{xy}, \qquad f^* = \phi_x \frac{|\nabla \phi|^2}{\phi}, \qquad (3.51)$$

$$g = -2 < \nabla \phi, \nabla \phi_y > +2\delta \, \phi_{yy}, \qquad g^* = (\phi_y - \delta) \frac{|\nabla \phi|^2}{\phi}, \qquad (3.52)$$

and $\delta \in \mathbb{R}$ will be suitably chosen subsequently. We notice that

$$f_y + g_x = 0$$

and therefore there exists a function $P = P(x, y)$ such that

$$f = P_x, \qquad -g = P_y.$$

This gives in particular

$$\Delta P = f_x - g_y. \qquad (3.53)$$

An easy calculation shows that

$$P = |\nabla \phi|^2 - 2\delta \, \phi_y. \qquad (3.54)$$

We obtain from (3.54)

$$\Delta P = 2||\nabla^2 \phi||^2 + 2 < \nabla \phi, \nabla(\Delta \phi) > -2\delta \, (\Delta \phi)_y, \qquad (3.55)$$

where we have denoted with $\nabla^2 \phi$ the Hessian matrix of ϕ. We now use (3.49) to compute ΔP. First, we see that

$$2 < \nabla \phi, \nabla(\Delta \phi) > = -(n+2) \frac{|\nabla \phi|^4}{\phi^2} + 2(n+2) \frac{< \nabla^2 \phi(\nabla \phi), \nabla \phi >}{\phi}$$

$$- \frac{2a}{x} < \nabla \phi, \nabla \phi_x > -\frac{2b}{y} < \nabla \phi, \nabla \phi_y > +\frac{2a}{x^2} \phi_x^2 + \frac{2b}{y^2} \phi_y^2 - \frac{n}{y^2} \phi_y. \qquad (3.56)$$

We also find

$$(\Delta\phi)_y = -\frac{n+2}{2}\frac{\phi_y|\nabla\phi|^2}{\phi^2} + (n+2)\frac{<\nabla\phi,\nabla\phi_y>}{\phi}$$
$$-\frac{a}{x}\phi_{xy} - \frac{b}{y}\phi_{yy} + \frac{b}{y^2}\phi_y - \frac{n}{2y^2}. \tag{3.57}$$

At this point we introduce the function

$$h = \gamma\,\phi^{-(n+1)}, \tag{3.58}$$

where $\gamma = \gamma(x,y)$ is a strictly positive function on Ω which will be determined subsequently. With F and G as in (3.50) we consider the differential expression

$$(hF)_x - (hG)_y = h(f_x - g_y) - h(f_x^* + g_y^*) + h_x F - h_y G$$
$$= h[\Delta P - (f_x^* + g_y^*) + \frac{n+1}{\phi}(\phi_y G - \phi_x F)]$$
$$+ \phi^{-(n+1)}(\gamma_x F - \gamma_y G). \tag{3.59}$$

A computation gives

$$f_x^* + g_y^* = \Delta\phi\frac{|\nabla\phi|^2}{\phi} - \frac{|\nabla\phi|^4}{\phi^2} + 2\frac{<\nabla^2\phi(\nabla\phi),\nabla\phi>}{\phi} \tag{3.60}$$
$$+ \delta\frac{\phi_y|\nabla\phi|^2}{\phi^2} - 2\delta\frac{<\nabla\phi,\nabla\phi_y>}{\phi},$$

$$\phi_y G - \phi_x F = \frac{|\nabla\phi|^4}{\phi} - 2 <\nabla^2\phi(\nabla\phi),\nabla\phi> +2\delta <\nabla\phi,\nabla\phi_y>$$
$$- \delta\frac{\phi_y|\nabla\phi|^2}{\phi}. \tag{3.61}$$

Using (3.60) and (3.61) we obtain from (3.59)

$$(hF)_x - (hG)_y = h\left[\Delta P - \Delta\phi\frac{|\nabla\phi|^2}{\phi} + (n+2)\frac{|\nabla\phi|^4}{\phi^2}\right.$$
$$- 2(n+2)\frac{<\nabla^2\phi(\nabla\phi),\nabla\phi>}{\phi} + 2\delta(n+2)\frac{<\nabla\phi,\nabla\phi_y>}{\phi}$$
$$\left.- \delta(n+2)\frac{\phi_y|\nabla\phi|^2}{\phi^2}\right] + \phi^{-(n+1)}(\gamma_x F - \gamma_y G). \tag{3.62}$$

At this point we substitute (3.55), (3.56) and (3.57) in (3.62) to obtain

$$
\begin{aligned}
(hF)_x &- (hG)_y \\
= h\Bigg\{ &\left(2\,||\nabla^2\phi||^2 - (\Delta\phi)^2\right) + (\Delta\phi)^2 - (n+2)\frac{|\nabla\phi|^4}{\phi^2} + 2(n+2)\frac{<\nabla^2\phi(\nabla\phi),\nabla\phi>}{\phi} \\
&- \frac{2a}{x} <\nabla\phi,\nabla\phi_x> - \frac{2b}{y} <\nabla\phi,\nabla\phi_y> + \frac{2a}{x^2}\,\phi_x^2 + \frac{2b}{y^2}\,\phi_y^2 - \frac{n}{y^2}\,\phi_y \\
&- 2\delta\left(-\frac{n+2}{2}\cdot\frac{\phi_y|\nabla\phi|^2}{\phi^2} + (n+2)\frac{<\nabla\phi,\nabla\phi_y>}{\phi}\right) \\
&- \frac{a}{x}\,\phi_{xy} - \frac{b}{y}\,\phi_{yy} + \frac{b}{y^2}\,\phi_y - \frac{n}{2y^2}\right) - \Delta\phi\frac{|\nabla\phi|^2}{\phi} + (n+2)\frac{|\nabla\phi|^4}{\phi^2} \\
&- 2(n+2)\frac{<\nabla^2\phi(\nabla\phi),\nabla\phi>}{\phi} + 2\delta(n+2)\frac{<\nabla\phi,\nabla\phi_y>}{\phi} \\
&- \delta(n+2)\frac{\phi_y|\nabla\phi|^2}{\phi^2}\Bigg\} + \phi^{-(n+1)}(\gamma_x F - \gamma_y G). \quad (3.63)
\end{aligned}
$$

The expression in (3.63) can be simplified as follows

$$
\begin{aligned}
(hF)_x - (hG)_y = h\Bigg[&\left(2\,||\nabla^2\phi||^2 - (\Delta\phi)^2\right) + \Delta\phi\left(\Delta\phi - \frac{|\nabla\phi|^2}{\phi}\right) \\
&- \frac{2a}{x} <\nabla\phi,\nabla\phi_x> - \frac{2b}{y} <\nabla\phi,\nabla\phi_y> + \frac{2a}{x^2}\phi_x^2 + \frac{2b}{y^2}\phi_y^2 - \frac{n}{y^2}\phi_y \\
&+ \frac{2\delta a}{x}\,\phi_{xy} + \frac{2\delta b}{y}\,\phi_{yy} - \frac{2\delta b}{y^2}\phi_y + \frac{\delta n}{y^2}\Bigg] + \phi^{-(n+1)}(\gamma_x F - \gamma_y G). \quad (3.64)
\end{aligned}
$$

Next we evaluate the expression

$$
\begin{aligned}
\Delta\phi\Bigg[&\Delta\phi - \frac{|\nabla\phi|^2}{\phi}\Bigg] - \frac{2a}{x} <\nabla\phi,\nabla\phi_x> - \frac{2b}{y} <\nabla\phi,\nabla\phi_y> \\
&+ \frac{2a}{x^2}\phi_x^2 + \frac{2b}{y^2}\phi_y^2 - \frac{n}{y^2}\,\phi_y + \frac{2\delta a}{x}\phi_{xy} + \frac{2\delta b}{y}\phi_{yy} - \frac{2\delta b}{y^2}\phi_y + \frac{\delta n}{y^2} \\
= &\frac{n(n+2)}{4}\frac{|\nabla\phi|^4}{\phi^2} - \frac{a(n+1)}{x}\frac{\phi_x|\nabla\phi|^2}{\phi} - \frac{b(n+1)}{y}\frac{\phi_y|\nabla\phi|^2}{\phi} \\
&+ \frac{n(n+1)}{2y}\frac{|\nabla\phi|^2}{\phi} + \frac{2ab}{xy}\phi_x\phi_y + \frac{a(a+2)}{x^2}\phi_x^2 + \frac{b(b+2)}{y^2}\phi_y^2 \\
&- \frac{an}{xy}\phi_x - \frac{bn+n+2\delta b}{y^2}\phi_y + \frac{n(n+4\delta)}{4y^2} \\
&- \frac{2a}{x} <\nabla\phi,\nabla\phi_x> - \frac{2b}{y} <\nabla\phi,\nabla\phi_y> + \frac{2\delta a}{x}\phi_{xy} + \frac{2\delta b}{y}\phi_{yy}. \quad (3.65)
\end{aligned}
$$

We now calculate

$$\gamma_x F - \gamma_y G = 2\left[\gamma_x < \nabla\phi, \nabla\phi_x > + \gamma_y < \nabla\phi, \nabla\phi_y >\right]$$
$$- 2\delta < \nabla\gamma, \nabla\phi_y > - < \nabla\gamma, \nabla\phi > \frac{|\nabla\phi|^2}{\phi} + \delta\gamma_y \frac{|\nabla\phi|^2}{\phi}. \quad (3.66)$$

The next step is to compute

$$\frac{n+2}{n}\left(\Delta\phi - \frac{|\nabla\phi|^2}{\phi}\right)^2 = \frac{n(n+2)}{4}\frac{|\nabla\phi|^4}{\phi^2} + a^2(1 + \frac{2}{n})\frac{\phi_x^2}{x^2} + b^2(1 + \frac{2}{n})\frac{\phi_y^2}{y^2}$$
$$+ \frac{n(n+2)}{4}\frac{1}{y^2} - \frac{a(n+2)}{x}\frac{\phi_x|\nabla\phi|^2}{\phi} - \frac{b(n+2)}{y}\frac{\phi_y|\nabla\phi|^2}{\phi}$$
$$+ \frac{n(n+2)}{2y}\frac{|\nabla\phi|^2}{\phi} + \frac{2ab(n+2)}{n}\frac{\phi_x\phi_y}{xy} - \frac{a(n+2)}{xy}\phi_x - \frac{b(n+2)}{y^2}\phi_y. \quad (3.67)$$

Subtracting (3.67) from (3.65) we find

$$\Delta\phi\left[\Delta\phi - \frac{|\nabla\phi|^2}{\phi}\right] - \frac{2a}{x} < \nabla\phi, \nabla\phi_x > - \frac{2b}{y} < \nabla\phi, \nabla\phi_y >$$
$$+ \frac{2a}{x^2}\phi_x^2 + \frac{2b}{y^2}\phi_y^2 - \frac{n}{y^2}\phi_y + \frac{2\delta a}{x}\phi_{xy} + \frac{2\delta b}{y}\phi_{yy} - \frac{2\delta b}{y^2}\phi_y + \frac{\delta n}{y^2}$$
$$- \frac{n+2}{n}\left(\Delta\phi - \frac{|\nabla\phi|^2}{\phi}\right)^2$$
$$= \frac{a}{x}\frac{\phi_x|\nabla\phi|^2}{\phi} + \frac{b}{y}\frac{\phi_y|\nabla\phi|^2}{\phi} - \frac{n}{2y}\frac{|\nabla\phi|^2}{\phi} - \frac{4ab}{n}\frac{\phi_x\phi_y}{xy} + \frac{2ab}{nx^2}\phi_x^2 + \frac{2ab}{ny^2}\phi_y^2 + \frac{2a}{xy}\phi_x$$
$$+ \frac{(2\delta-1)n}{2y^2} - \frac{2a}{x} < \nabla\phi, \nabla\phi_x > - \frac{2b}{y} < \nabla\phi, \nabla\phi_y >$$
$$+ \frac{2\delta a}{x}\phi_{xy} + \frac{2\delta b}{y}\phi_{yy} + \frac{b(1-2\delta)-a}{y^2}\phi_y. \quad (3.68)$$

We now multiply equation (3.66) by γ^{-1} and add it to (3.68) obtaining

$$E \stackrel{\text{def}}{=} \Delta\phi \left[\Delta\phi - \frac{|\nabla\phi|^2}{\phi} \right] - \frac{2a}{x} <\nabla\phi, \nabla\phi_x> - \frac{2b}{y} <\nabla\phi, \nabla\phi_y>$$

$$+ \frac{2a}{x^2} \phi_x^2 + \frac{2b}{y^2} \phi_y^2 - \frac{n}{y^2} \phi_y + \frac{2\delta a}{x} \phi_{xy} + \frac{2\delta b}{y} \phi_{yy} - \frac{2\delta b}{y^2} \phi_y + \frac{\delta n}{y^2}$$

$$- \frac{n+2}{n} \left(\Delta\phi - \frac{|\nabla\phi|^2}{\phi} \right)^2 + \gamma^{-1} \left(\gamma_x F - \gamma_y G \right)$$

$$= \frac{a}{x} \frac{\phi_x |\nabla\psi|^2}{\phi} + \frac{b}{y} \frac{\phi_y |\nabla\phi|^2}{\phi} - \frac{n}{2y} \frac{|\nabla\phi|^2}{\phi} - \frac{4ab}{n} \frac{\phi_x\phi_y}{xy} + \frac{2ab}{nx^2} \phi_x^2 + \frac{2ab}{ny^2} \phi_y^2$$

$$+ \frac{2a}{xy} \phi_x + \frac{(2\delta-1)n}{2y^2} - \frac{2a}{x} <\nabla\phi, \nabla\phi_x> - \frac{2b}{y} <\nabla\phi, \nabla\phi_y>$$

$$+ \frac{2\delta a}{x} \phi_{xy} + \frac{2\delta b}{y} \phi_{yy} + \frac{b(1-2\delta)-a}{y^2} \phi_y$$

$$+ \gamma^{-1} \left\{ 2 \left[\gamma_x <\nabla\phi, \nabla\phi_x> + \gamma_y <\nabla\phi, \nabla\phi_y> \right] \right.$$

$$\left. - 2\delta <\nabla\gamma, \nabla\phi_y> - <\nabla\gamma, \nabla\phi> \frac{|\nabla\phi|^2}{\phi} + \delta\gamma_y \frac{|\nabla\phi|^2}{\phi} \right\}. \quad (3.69)$$

At this point we make a suitable choice of the function γ. We let

$$\gamma(x,y) = x^a y^b, \quad (3.70)$$

with which the last term in (3.69) takes the form

$$\gamma^{-1} \left\{ 2 \left[\gamma_x <\nabla\phi, \nabla\phi_x> + \gamma_y <\nabla\phi, \nabla\phi_y> \right] \right.$$

$$\left. - <\nabla\gamma, \nabla\phi_y> - <\nabla\gamma, \nabla\phi> \frac{|\nabla\phi|^2}{\phi} + \frac{\gamma_y}{2} \frac{|\nabla\phi|^2}{\phi} \right\}$$

$$= \frac{2a}{x} <\nabla\phi, \nabla\phi_x> + \frac{2b}{y} <\nabla\phi, \nabla\phi_y> - \frac{2\delta a}{x} \phi_{xy} - \frac{2\delta b}{y} \phi_{yy}$$

$$- \frac{a}{x} \frac{\phi_x |\nabla\phi|^2}{\phi} - \frac{b}{y} \frac{\phi_y |\nabla\phi|^2}{\phi} + \frac{\delta b}{y} \frac{|\nabla\phi|^2}{\phi}. \quad (3.71)$$

A substitution of (3.71) in (3.69) gives

$$E = \frac{2\delta b - n}{2y} \frac{|\nabla\phi|^2}{\phi} + \frac{2ab}{nx^2} \phi_x^2 + \frac{2ab}{ny^2} \phi_y^2 + \frac{2a}{xy} \phi_x \quad (3.72)$$

$$- \frac{4ab}{n} \frac{\phi_x\phi_y}{xy} + \frac{(2\delta-1)n}{2y^2} + \frac{b(1-2\delta)-a}{y^2} \phi_y.$$

We finally choose δ in (3.72) as follows

$$\delta = \frac{n}{2b}. \quad (3.73)$$

With this choice we obtain from (3.72)

$$E = \frac{2ab}{nx^2}\,\phi_x^2 + \frac{2ab}{ny^2}\,\phi_y^2 - \frac{4ab}{n}\,\frac{\phi_x\phi_y}{xy} + \frac{2a}{xy}\,\phi_x - \frac{2a}{y^2}\,\phi_y + \frac{an}{2by^2}$$

$$= \frac{2ab}{n}\left(\frac{\phi_x}{x} - (\frac{\phi_y}{y} - \frac{n}{2by})\right)^2. \quad (3.74)$$

Summarizing, we have proved the following identity.

Theorem 3.4.2. *Let ϕ be a positive solution to the equation (3.49) in $\Omega = \{(x,y) \in \mathbb{R}^2 \mid x > 0, y > 0\}$. With $h = x^a y^b \phi^{-(n+1)}$, and F and G as in (3.50), the following identity holds*

$$(hF)_x - (hG)_y = h\left\{\left[2\,\|\nabla^2\phi\|^2 - (\Delta\phi)^2\right] + \frac{n+2}{n}\left(\Delta\phi - \frac{|\nabla\phi|^2}{\phi}\right)^2 \right.$$

$$\left. + \frac{2ab}{n}\left(\frac{\phi_x}{x} - (\frac{\phi_y}{y} - \frac{n}{2by})\right)^2\right\}.$$

Before proceeding we note explicitly that thanks to Schwarz' inequality the term within square brackets in the right-hand side of the above identity is non-negative, thus the right-hand side is the sum of three non-negative terms. Our next step is to use the Kelvin transform to obtain the exact asymptotic behavior of the function ϕ.

Lemma 3.4.3. *Let $u \not\equiv 0$ be an entire solution to (3.1) in a group of Iwasawa type $\boldsymbol{G}$. One has $u > 0$ in $\boldsymbol{G}$ and $u \in C^\infty(\boldsymbol{G})$. Suppose in addition that u is cylindrically symmetric, let $\Phi = u^{-4/(Q-2)}$ and denote by ϕ the symmetric part of Φ as in (3.43). One has for some constant $C = C(\boldsymbol{G}) > 0$ and large enough $z = (x,y) \in \Omega$*

$$C^{-1}|z|^2 \le \phi(z) \le C|z|^2, \qquad |\nabla\phi(z)| \le C|z|, \qquad |\nabla^2\phi(z)| \le C. \quad (3.75)$$

Proof. The proof is a simple consequence of the properties of the Kelvin transform in a group of Iwasawa type. Let u^* be the Kelvin transform of u. An easy computation, using (2.12) and (3.46), gives $N(g) = 2|z|^{1/2}$. From (3.46) and (2.33) we find also

$$|\eta_1| = \frac{y^{1/2}}{2|z|}, \qquad |\eta_2| = \frac{x}{16|z|^2}. \quad (3.76)$$

Notice that when $z \to \infty$ we have $|\eta_1|, |\eta_2| \to 0$. Since the Kelvin transform is an involution, an argument very similar to that in the end of the proof of Theorem 3.3.1 gives the asymptotic for u in (3.75). In fact, we see that both u and u^* are entire solutions to (3.1) and they have the decay (3.22). Using (3.46) again we obtain (3.75) for both u and u^*. The bounds for the derivatives follow from the homogeneity of the arguments of

$$\phi(x,y) = |z|^2 \phi^*\left(\frac{y^{1/2}}{2|z|}, \frac{x}{16|z|^2}\right), \tag{3.77}$$

where $\phi^* = (u^*)^{-\frac{4}{Q-2}}$ and from differentiation. $\qquad\square$

We turn to the proof of Theorem 3.4.1.

Proof. We recall that we are assuming $dim V_2 = k \geq 2$, so that $a \geq 1$, and therefore $h \equiv 0$ on $\partial\Omega$. We consider the functions Φ and ϕ as in Theorem 3.4.2 and Lemma 3.4.3. For every $R > 0$ set $\Omega_R = \Omega \cap B(0,R)$, $\Gamma_R = \Omega \cap \partial B(0,R)$. Integrating the left-hand side of the identity in Theorem 3.4.2 we find

$$\int_{\Omega_R} [(hF)_x - (hG)_y]\, dxdy = \frac{1}{R}\int_{\Gamma_R} h\,[xF - yG]\, ds. \tag{3.78}$$

We now use (3.50), (3.51), (3.52) and Lemma 3.4.3 to infer

$$\left|\frac{1}{R}\int_{\Gamma_R} h\,[xF - yG]\, ds\right| \leq C\,R^{-n} \to 0 \qquad \text{as} \quad R \to \infty. \tag{3.79}$$

Combining (3.78) with (3.79) and with Theorem 3.4.2, we finally obtain

$$\int_\Omega h\left\{\left[2\,||\nabla^2\phi||^2 - (\Delta\phi)^2\right] + \frac{n+2}{n}\left(\Delta\phi - \frac{|\nabla\phi|^2}{\phi}\right)^2\right.$$
$$\left. + \frac{2ab}{n}\left(\frac{\phi_x}{x} - (\frac{\phi_y}{y} - \frac{n}{2by})\right)^2\right\} dxdy = 0.$$

The latter equation implies

$$2\,||\nabla^2\phi||^2 = (\Delta\phi)^2, \quad \Delta\phi - \frac{|\nabla\phi|^2}{\phi} = 0, \quad \frac{\phi_x}{x} = \frac{\phi_y}{y} - \frac{n}{2by}. \tag{3.80}$$

From the first two equations in (3.80) and from Lemma 3.4.3 we conclude in a classical fashion (see, e.g., [169] or also [102]) that ϕ must be of the type

$$\phi(x,y) = A^2\,(x^2 + y^2) + 2A\alpha x + 2B\beta y + \alpha^2 + \beta^2 \tag{3.81}$$

for some numbers A, B, α and β, with $A^2 = B^2$. On the other hand, the third equation in (3.80) implies that must be

$$\alpha = 0 \qquad \text{and} \qquad \beta = \frac{n}{4bB}.$$

Recalling that $x = |\xi_2|, y = |\xi_1|^2/4$ one easily concludes from the above that

$$\phi(|\xi_1|, |\xi_2|) = \frac{A^2}{16} \left[(\frac{a+b}{bA^2} + |\xi_1|^2)^2 + 16|\xi_2|^2 \right] \qquad (3.82)$$

for some $A \neq 0$. Using (3.48) we can rewrite (3.82) as follows

$$\phi(|\xi_1|, |\xi_2|) = \frac{Q-2}{16m\epsilon^2} \left[(\epsilon^2 + |\xi_1|^2)^2 + 16|\xi_2|^2 \right] \qquad (3.83)$$

where $\epsilon^2 = \frac{Q-2}{mA}$. Finally, keeping in mind that $\phi = v^{-4/(Q-2)}$, and that $u = (1/\lambda)v$, with λ given by (3.40), we obtain

$$u(g) = C_\epsilon \left((\epsilon^2 + |x(g)|^2)^2 + 16|y(g)|^2 \right)^{-(Q-2)/4},$$

with $C_\epsilon = [m(Q-2)\epsilon^2]^{(Q-2)/4}$. All other cylindrically symmetric solutions are obtained from this one by left-translation. This completes the proof. $\square$

3.5 Solution of the partially symmetric Yamabe problem

We are now ready to prove the main result of this chapter.

Theorem 3.5.1. *Let G be a group of Iwasawa type. If $U \not\equiv 0$ is an entire solution to (3.1) having partial symmetry, then up to group translations we must have $u = K_\epsilon$, for some $\epsilon > 0$, where K_ϵ is the function in Theorem 2.4.1 and u is as in (3.2).*

Theorem 3.5.1 is a direct consequence of Theorem 3.3.1 and Theorem 3.4.1.

As a consequence of Theorem 3.5.1 we obtain the following result. In the sequel we denote by $\overset{o}{X}_{ps}(G)$ the subset of $\overset{o}{\mathcal{D}}^{1,2}(G)$ of the functions having partial symmetry.

Theorem 3.5.2. *Let G be a group of Iwasawa type. Consider the restriction to $\overset{o}{X}_{ps}(G)$ of the embedding of $\overset{o}{\mathcal{D}}^{1,2}(G)$ into $L^{2Q/(Q-2)}(G)$. For every $u \in \overset{o}{X}_{ps}(G)$ one has*

$$\left(\int_G |u|^{2^*} dH(g) \right)^{1/2^*} \leq S_2^{ps} \left(\int_G |Xu|^2 \, dH(g) \right)^{1/2},$$

with

$$S_2^{ps} = \frac{1}{\sqrt{m(m+2(k-1))}} \, 4^{\frac{k}{m+2k}} \pi^{-\frac{m+k}{2(m+2k)}} \left(\frac{\Gamma(m+k)}{\Gamma\left(\frac{m+k}{2}\right)}\right)^{\frac{1}{m+2k}}. \qquad (3.84)$$

An extremal is given by the function

$$f(g) = \gamma(m,k) \left[(1+|x(g)|^2)^2 + 16|y(g)|^2\right]^{-(Q-2)/4},$$

where

$$\gamma(m,k) = \left[4^k \, \pi^{-(m+k)/2(m+2k)} \, \frac{\Gamma(m+k)}{\Gamma((m+k)/2)}\right]^{(m+2(k-1))/2(m+2k)}.$$

Any other non-negative extremal is obtained from f by (1.28) and (1.29).

Proof. Let $\overset{o}{\mathcal{D}}_{ps}^{1,2}(G)$ denote the subspace of $\overset{o}{\mathcal{D}}^{1,2}(G)$ of the functions U such that

$$U(g) = u(|x(g)|, y(g)),$$

for some function $u : [0, \infty) \times \mathbb{R}^k \to \mathbb{R}$. We start with the observation that we can restrict our considerations to the non-negative functions in $\overset{o}{\mathcal{D}}_{ps}^{1,2}(G)$, i.e.,

$$I_{ps} \overset{def}{=} \inf\left\{\int_G |Xu|^2 dH : u \in \overset{o}{\mathcal{D}}_{ps}^{1,2}(G), u \geq 0, \int_G |u|^{2^*} dH = 1\right\}, \qquad (3.85)$$

where $I_{ps} \overset{def}{=} (S_2^{ps})^{-2}$. This follows from the invariance of the integrals under left translation, and the fact that if $U \in \overset{o}{\mathcal{D}}^{1,2}(G)$, then also $|U| \in \overset{o}{\mathcal{D}}^{1,2}(G)$ and $|XU| = |X|U||$ for a.e. $g \in G$. From Theorem 1.5.2 the inf in (3.85) is achieved. Let $v \in \overset{o}{\mathcal{D}}_{ps}^{1,2}(G)$ be a function for which the inf is attained, thus

$$I_{ps} = \int_G |Xv|^2 \, dH, \qquad \int_G v^{2^*} \, dH = 1.$$

Writing the Euler-Lagrange equation of the constrained problem (3.85) we see that v is a positive entire solution of $\mathcal{L}v = -I_{ps} \, v^{(Q+2)/(Q-2)}$. Let $u \overset{def}{=} I_{ps}^{\frac{1}{2^*-2}} v$, then u is a positive entire solution of (3.1). Since $u \in \overset{o}{\mathcal{D}}_{ps}^{1,2}(G)$, Theorem 3.5.1 shows that u, modulo translations in the center, belongs to the one-parameter family of positive entire solutions, namely the functions K_ϵ in Theorem 2.4.1. From the definition of u, it is easy to see that

$$I_{ps} = \left(\int_G |Xu|^2 dH \right)^{\frac{2}{Q}}.$$

Since u is a positive entire solution of (3.1) we have $\int_G |Xu|^2 \, dH(g) = \int_G u^{2^*} \, dH(g)$, which shows that $I_{ps} = \|u\|_{L^{2^*}(G)}^{2^*-2}$. Note that $K_\epsilon = \delta_{1/\epsilon}K$, where we have let $K = K_1$, and an easy computation gives $\|\delta_{1/\epsilon}K\|_{L^{2^*}(G)} = \|K\|_{L^{2^*}(G)}$. As already remarked, all considered integrals are invariant under the translations (1.28) as well. From the above considerations we infer

$$I_{ps} = \left[(m(Q-2))^{Q/2} \int_G \frac{1}{[(1+|x(g)|^2)^2 + 16|y(g)|^2]^{Q/2}} \, dH(g) \right]^{2/Q}.$$
$$(3.86)$$

To obtain the best constant S_2 at this point we are left with the computation of the integral in the right-hand side of (3.86). If $U(g) = u(|x(g)|, |y(g)|) \in L^1(G)$ is a function with cylindrical symmetry on G we have

$$\int_G U(g)dH(g) = \int_{\mathbb{R}^m \times \mathbb{R}^k} u(|x|, |y|)dxdy. \qquad (3.87)$$

Using (3.87) we find

$$\int_G \frac{1}{[(1+|x(g)|^2)^2 + 16|y(g)|^2]^{Q/2}} \, dH(g)$$
$$= \int_{\mathbb{R}^m \times \mathbb{R}^k} \frac{dxdy}{[(1+|x|^2)^2 + 16|y|^2]^{Q/2}}$$
$$= 4^{-k} \int_{\mathbb{R}^m} \frac{dx}{(1+|x|^2)^{Q-k}} \int_{\mathbb{R}^k} \frac{dy}{(1+|y|^2)^{Q/2}}. \qquad (3.88)$$

Consider now the integral

$$\int_{\mathbb{R}^n} \frac{dt}{(1+|t|^2)^a} = \frac{\sigma_n}{2} B\left(\frac{n}{2}, a - \frac{n}{2} \right), \qquad a > \frac{n}{2}, \qquad (3.89)$$

where σ_n denotes the $(n-1)$-dimensional measure of the Euclidean unit sphere in $\mathbb{R}^n$, and $B(x, y)$ is the Beta function. Recalling the two formulas for the area of the unit sphere and the beta function

$$\sigma_n = \frac{2\pi^{n/2}}{\Gamma(n/2)} \quad \text{and} \quad B(x, y) = \frac{\Gamma(x)\,\Gamma(y)}{\Gamma(x+y)},$$

where Γ indicates Euler's Gamma function, we conclude from (3.89)

$$\int_{\mathbb{R}^n} \frac{dt}{(1+|t|^2)^a} = \pi^{n/2}\frac{\Gamma(a-\frac{n}{2})}{\Gamma(a)}. \tag{3.90}$$

Using (3.90) in (3.88), and recalling that $Q = m + 2k$, we finally obtain

$$\int_G \frac{1}{[(1+|x(g)|^2)^2 + 16|y(g)|^2]^{Q/2}}\, dH(g) = 4^{-k}\pi^{(m+k)/2}\frac{\Gamma(\frac{m+k}{2})}{\Gamma(m+k)}. \tag{3.91}$$

Substitution of (3.91) into (3.86) gives

$$I = m(Q-2)\left(4^{-k}\,\pi^{(m+k)/2}\frac{\Gamma(\frac{m+k}{2})}{\Gamma(m+k)}\right)^{2/Q}.$$

Therefore,

$$S_2^{ps} = \frac{1}{\sqrt{m(m+2(k-1))}}\, 4^{\frac{k}{m+2k}}\,\pi^{-\frac{m+k}{2m+4k}}\,\left(\frac{\Gamma(m+k)}{\Gamma(\frac{m+k}{2})}\right)^{\frac{1}{m+2k}}. \tag{3.92}$$

This completes the proof of Theorem 3.5.2. $\square$

We note explicitly that the above precise value of the best constant is valid under the assumption that the horizontal gradient is taken with respect to an orthonormal basis $X_1, \dots, X_m$ of the first layer for which the group is of H-type, cf. (2.10) and the paragraph above it.

Remark 3.5.3. There is the conjecture made after [[75], Theorem 1.1] that $S_2^{ps} = S_2$. This is confirmed in all but the octonian case due to the results of G. Talenti [159] and Th. Aubin [14] and [16], D. Jerison and J. Lee [103] - see also [70], and [96] - see Theorem 6.1.1 and Remark 6.7.5.

3.6   Applications. Euclidean Hardy-Sobolev inequalities

The goal of this section is to determine the best constant in a Hardy-Sobolev embedding theorem involving the distance to a subspace of $\mathbb{R}^n$.

Let $n \geq 3$ and $2 \leq k \leq n$. For a point z in $\mathbb{R}^m = \mathbb{R}^k \times \mathbb{R}^{n-k}$ we shall write $z = (x, y)$, where $x \in \mathbb{R}^k$ and $y \in \mathbb{R}^{n-k}$. The following Hardy-Sobolev inequality was proven in Theorem 2.1 of [18].

Theorem 3.6.1 ([18]). *Let $n \geq 3$, $2 \leq k \leq n$, and p, s be real numbers satisfying $1 < p < n$, $0 \leq s \leq p$, and $s < k$. There exists a positive constant $S_{p,s} = S(s, p, n, k)$ such that for all $u \in D^{1,p}(\mathbb{R}^m)$ we have*

$$\left(\int_{\mathbb{R}^m} \frac{|u|^{\frac{p(n-s)}{n-p}}}{|x|^s}\, dz\right)^{\frac{n-p}{p(n-s)}} \leq S_{p,s}\left(\int_{\mathbb{R}^m} |\nabla u|^p\, dz\right)^{\frac{1}{p}}. \tag{3.93}$$

When $k = n$ the above inequality becomes the Caffarelli-Kohn-Nirenberg inequality, see [34], for which the optimal constant $S_{p,s}$ was found in [77]. The case $p = 2$ was considered earlier in [134] and [81]. When $p = 2$ and $k = n$ the sharp constant was computed in [81].

The main result of this section is the proof of the following theorem in which we find the extremals and the best constant in (3.93) when $p = 2$, $s = 1$ and $2 \le k \le n$ in Theorem 3.6.1.

Theorem 3.6.2. *Suppose $n \ge 3$ and $2 \le k \le n$. There exists a positive constant $K = K_{n,k,2}$ such that for all $u \in D^{1,2}(\mathbb{R}^m)$ we have*

$$\left(\int_{\mathbb{R}^{n-k}} \int_{\mathbb{R}^k} \frac{|u|^{\frac{2(n-1)}{n-2}}}{|x|} \, dxdy \right)^{\frac{n-2}{2(n-1)}} \le K \left(\int_{\mathbb{R}^m} |\nabla u|^2 \, dz \right)^{\frac{1}{2}}. \tag{3.94}$$

Furthermore, K is given in (3.105) and the positive extremals are the functions

$$v = \lambda^{-(n-2)} \left(\frac{4}{(n-2)^2} \right)^{-\frac{n-2}{2}} K^{-(n-1)} \left((|x| + \frac{n-2}{4a\lambda^2})^2 + |y - y_o|^2 \right)^{-\frac{n-2}{2}}, \tag{3.95}$$

where $\lambda > 0$, $y_o \in \mathbb{R}^{n-k}$.

In [127] the authors obtain independently the above theorem proving identities following the lines of [75]. In the proof below, [167], we show the result as a direct consequence of Theorem 3.4.1 by relating extremals on the Heisenberg groups to extremals in the Euclidean setting.

3.6.1 A non-linear equation in $\mathbb{R}^n$ related to the Yamabe equation on groups of Heisenberg type

Suppose a and b are two natural numbers, $\lambda > 0$, and for $x, y \in \mathbb{R}^+ = (0, +\infty)$, define the function

$$\phi = \lambda^2 \left[(x + \alpha)^2 + (y + \beta)^2 \right],$$

where $\alpha, \beta \in \mathbb{R}$.

Proposition 3.6.3. *The function ϕ satisfies the following equation in the plane*

$$\Delta \phi - \frac{a + b + 2}{2} \frac{|\nabla \phi|^2}{\phi} + \frac{a}{x} \phi_x + \frac{b}{y} \phi_y = \frac{2a\lambda^2 \alpha}{x} + \frac{2b\lambda^2 \beta}{y}, \quad xy \neq 0. \tag{3.96}$$

Proof. Set $\xi = \lambda(x + \alpha)$, $\eta = \lambda(y + \beta)$ and define $\tilde{\phi}(\xi, \eta) = \phi(x, y)$. From $\frac{\partial}{\partial x} = \lambda \frac{\partial}{\partial \xi}$ and $\frac{\partial}{\partial y} = \lambda \frac{\partial}{\partial \eta}$ we have

$$\Sigma \stackrel{def}{=} \Delta\phi - \frac{a+b+2}{2}\frac{|\nabla\phi|^2}{\phi} + \frac{a}{x}\phi_x + \frac{b}{y}\phi_y$$

$$= \lambda\Delta\tilde{\phi} - \frac{a+b+2}{2}\lambda^2\frac{|\nabla\tilde{\phi}|^2}{\tilde{\phi}} + \frac{a\lambda}{x}\tilde{\phi}_\xi + \frac{b\lambda}{y}\tilde{\phi}_\eta.$$

Since $\tilde{\phi} = \xi^2 + \eta^2$ we have

$$\Sigma = 4\lambda^2 - \frac{n+2}{2}\lambda^2\frac{4(\xi^2+\eta^2)}{\xi^2+\eta^2} + \frac{a}{x}2\lambda\xi + \frac{b}{y}2\lambda\eta$$

$$= -2n\lambda^2 + 2a\lambda^2 + 2b\lambda^2 + \frac{2a\lambda^2\alpha}{x} + \frac{2b\lambda^2\beta}{y}.$$

Hence, taking into account $a + b = n$, we proved $\Sigma = \frac{2a\lambda^2\alpha}{x} + \frac{2b\lambda^2\beta}{y}$. $\square$

Noting that $\Delta\phi + \frac{a}{x}\phi_x + \frac{b}{y}\phi_y$ is the Laplacian in $\mathbb{R}^n \equiv \mathbb{R}^{a+1} \times \mathbb{R}^{b+1}$, $n = a+b+2$, acting on functions with cylindrical symmetry, i.e., depending on $|\boldsymbol{x}|$ and $|\boldsymbol{y}|$ only, we are lead to the following question. Given two real numbers p_o and q_o find all positive solutions of the equation

$$\Delta u - \frac{n}{2}\frac{|\nabla u|^2}{u} = \frac{p_o}{|\boldsymbol{x}|} + \frac{q_o}{|\boldsymbol{y}|}, \quad (\boldsymbol{x}, \boldsymbol{y}) \in \mathbb{R}^n \equiv \mathbb{R}^{a+1} \times \mathbb{R}^{b+1},$$

which have at most a quadratic growth condition at infinity, $u \leq C(|\boldsymbol{x}|^2 + |\boldsymbol{y}|^2)$.

As usual a simple transformation allows to remove the appearance of the gradient in the above equation. For a function F we have $\Delta F(u) = F''(u)|\nabla u|^2 + F'(u)\Delta u$ and thus

$$\Delta u^\tau = \tau(\tau-1)u^{\tau-2}|\nabla u|^2 + \tau u^{\tau-1}\Delta u = \frac{\tau}{2}u^{\tau-2}(2u\Delta u + 2(\tau-1)|\nabla u|^2).$$

Therefore we choose τ such that $2(\tau - 1) = -n$, i.e., $\tau = \frac{2-n}{2}$ and then rewrite the equation for u as

$$\Delta u^{\frac{2-n}{2}} = \frac{2-n}{2}u^{\frac{2-n}{2}-2}(2u\Delta u - n|\nabla u|^2) = -\frac{(n-2)}{2}u^{\frac{2-n}{2}-1}\left(\frac{p_o}{|\boldsymbol{x}|} + \frac{q_o}{|\boldsymbol{y}|}\right).$$

This is the equation which we will study. As a consequence of the above calculations we can write a three parameter family of explicit solutions.

Proposition 3.6.4. *Let $\lambda > 0$. The function $v(\boldsymbol{x}, \boldsymbol{y})$ defined in $\mathbb{R}^n \equiv \mathbb{R}^{a+1} \times \mathbb{R}^{b+1}$ by the formula*

$$v(\boldsymbol{x}, \boldsymbol{y}) = \lambda^{2-n}\left((|\boldsymbol{x}|+\alpha)^2 + (|\boldsymbol{y}|+\beta)^2\right)^{\frac{2-n}{2}}, \quad (\boldsymbol{x}, \boldsymbol{y}) \in \mathbb{R}^n \equiv \mathbb{R}^{a+1} \times \mathbb{R}^{b+1},$$

$$\tag{3.97}$$

satisfies the equation

$$\Delta v = -v^{\frac{n}{n-2}} \left(\frac{p}{|x|} + \frac{q}{|y|} \right), \tag{3.98}$$

where

$$p = \alpha \, (n-2)\lambda^2 \, a, \qquad q = \beta \, (n-2)\lambda^2 \, b. \tag{3.99}$$

Let us observe that the above equation is invariant under rotations in the x or y variables. Also if v is a solution then a simple calculations shows that for any $t \neq 0$ the function $v_t(x, y) = t^{(n-2)/2} \, v(tx, ty)$ is also a solution.

Another observation is that the same principle works if we split $\mathbb{R}^n$ in more than two subspaces. For example, if we take three subspaces we can consider the equation

$$\Delta v = v^{\frac{n}{n-2}} f(|x|, |y|, |z|), \qquad f(x, y, z) = \frac{p}{|x|} + \frac{q}{|y|} + \frac{r}{|z|}$$

and ask the question of finding all positive solutions with the same behavior at infinity as the fundamental solution. Clearly the function

$$v = \lambda^{2-n} \left(\, (|x| + \alpha)^2 + (|y| + \beta)^2 + (|z| + \gamma)^2 \, \right)^{\frac{2-n}{2}},$$

with the obvious choice of α, β and γ is a solution.

3.6.2 *The best constant and extremals of the Hardy-Sobolev inequality*

In this Section we give the proof of Theorem 3.6.2. By Theorems 2.1 and 2.5 of [18] there is a constant K for which (3.94) holds and this constant is achieved, i.e., the equality is achieved. A non-negative extremal u of the naturally associated variational problem $\inf \int_{\mathbb{R}^m} |\nabla u|^2 \, dz$ subject to the constraint

$$\int_{\mathbb{R}^{n-k}} \int_{\mathbb{R}^k} \frac{|u|^{\frac{2(n-1)}{n-2}}}{|x|} \, dx dy = 1 \tag{3.100}$$

satisfies the Euler-Lagrange equation

$$\Delta u = -\frac{\Lambda}{|x|} u^{\frac{n}{n-2}}, \qquad u \in D^{1,2}(\mathbb{R}^m), \tag{3.101}$$

where $\Lambda = K^{\frac{2(n-1)}{n-2}}$. It can be seen [167] and [127] that u is a C^{∞} function on $|x| \neq 0$. Furthermore, $\nabla u \in L^{\infty}_{\text{loc}}(\mathbb{R}^n)$ and u is C^{∞} smooth in the y variables. In particular $u \in C^{0,\alpha}_{\text{loc}}(\mathbb{R}^n)$ for any $0 < \alpha < 1$.

It was proven in [150] that there are extremals with cylindrical symmetry, i.e., functions depending only on $|x|$ and $|y|$ for which the inequality becomes equality. On the other hand, it was shown in [129] that all extremals of inequality (3.93) when $p = 2$ have cylindrical symmetry after a suitable translation in the y variable, see also [41] and [124] for some related results. Thus, if $u \in \overset{o}{\mathcal{D}}{}^{1,2}(\mathbb{R}^n)$ is a function for which equality holds in (3.93) then

  i) for any $y \in \mathbb{R}^{n-k}$ the function $u(., y)$ is a radially symmetric decreasing function in $\mathbb{R}^k$;
  ii) there exists a $y_o \in \mathbb{R}^{n-k}$ such that for all $x \in \mathbb{R}^k$ the function $u(x, . + y_o)$ is a radially symmetric decreasing function on $\mathbb{R}^{n-k}$.

Thus, by performing a translation if necessary, we can assume that u has cylindrical symmetry. Introducing $\rho = |x|$, $r = |y|$ we have that u is a function of ρ and r. We define $U(\rho, r) = u$ by restricting u to two lines through the origin-one in $\mathbb{R}^k$, the other in $\mathbb{R}^{n-k}$. From the regularity of u it follows that U is a smooth function of r for any fixed ρ. For any fixed r it is a smooth function of ρ when $\rho \neq 0$, and Lipschitz for any ρ. Furthermore, in the first quadrant $\rho > 0, r > 0$ of the ρr-plane it satisfies the equation

$$\Delta U = -\frac{\Lambda}{\rho} \, U^{\frac{n}{n-2}} . \qquad (3.102)$$

Using the equation and the smoothness of U in r it is not hard to see that U has bounded first and second order derivatives on $((0,1) \times (0,1))$, cf. Lemma 3.6.5.

Let $\phi(\rho, r) = U^{-\frac{2}{n-2}}$. The calculations of Section 3.6.1 show that ϕ satisfies the following equation in the plane

$$\Delta\phi - \frac{n}{2}\frac{|\nabla\phi|^2}{\phi} + \frac{a}{\rho}\,\phi_\rho + \frac{b}{r}\,\phi_r - \frac{2\Lambda}{n-2}\frac{1}{\rho} = 0, \qquad (3.103)$$

where $a = k - 1$, $b = n - k - 1$. Let $\mu > 0$ and consider $\tilde{\phi} = \mu^{-1}\phi$. Clearly $\tilde{\phi}$ is a solution of

$$\Delta\tilde{\phi} - \frac{n}{2}\frac{|\nabla\tilde{\phi}|^2}{\tilde{\phi}} + \frac{a}{\rho}\,\tilde{\phi}_\rho + \frac{b}{r}\,\tilde{\phi}_r - \frac{2\Lambda}{\mu(n-2)}\frac{1}{\rho} = 0.$$

Let us choose μ such that $\frac{2\Lambda}{\mu(n-2)} = \frac{n-2}{2}$, i.e., $\mu = \frac{4\Lambda}{(n-2)^2}$. With this choice of μ we see that $\tilde{\phi}$ satisfies equation (3.49). Moreover, a small argument using the homogeneity of the Kelvin transform shows it satisfies the asymptotic behavior (3.75), except the inequality for the derivatives hold only on $|x| \neq 0$. We can apply the divergence formula (3.78) by noticing that the integrals on the ρ and r axes vanish as U has bounded first and second order derivatives in the punctured neighborhood of any point from the closed first quadrant, a fact which we observed above. Hence (3.81) after setting $|A| = \lambda$ gives $\tilde{\phi} = \lambda^2 \left[(r + \frac{n-2}{4a\lambda^2})^2 + s^2 \right]$. Recalling that $\phi = \mu \tilde{\phi}$ and the value of μ we come to

$$\phi = \lambda^2 \frac{4\Lambda}{(n-2)^2} \left[(r + \frac{n-2}{4a\lambda^2})^2 + s^2 \right].$$

This shows that v must equal

$$v = \lambda^{-(n-2)} \left(\frac{4}{(n-2)^2} \right)^{-\frac{n-2}{2}} \Lambda^{-\frac{n-2}{2}} \left[(|x| + \frac{n-2}{4a\lambda^2})^2 + |y|^2 \right]^{-\frac{n-2}{2}}$$

$$= \lambda^{-(n-2)} \left(\frac{n-2}{2} \right)^{n-2} K^{-(n-1)} \left[(|x| + \frac{n-2}{4a\lambda^2})^2 + |y|^2 \right]^{-\frac{n-2}{2}}.$$

The value of K is determined by (3.100) after fixing λ arbitrarily, say $\lambda = 1$, since the value of the integral in (3.100) is independent of λ. With this goal in mind we set $p = \frac{n-2}{4a}$ and note the identity

$$1 = \int_{\mathbb{R}^{n-k}} \int_{\mathbb{R}^k} \frac{1}{|x|} \left[\left(\frac{n-2}{2} \right)^{n-2} \frac{1}{K^{n-1}} \frac{1}{[(|x| + p)^2 + |y|^2]^{\frac{n-2}{2}}} \right]^{\frac{2(n-1)}{n-2}} dx dy$$

$$= \frac{1}{K^{\frac{2(n-1)^2}{n-2}}} \left(\frac{n-2}{2} \right)^{2(n-1)} \int_{\mathbb{R}^{n-k}} \int_{\mathbb{R}^k} \frac{1}{|x|} \frac{1}{[(|x| + p)^2 + |y|^2]^{n-1}} dx dy.$$

$$(3.104)$$

Let $a = |x| + p$. Then we compute

$$\int_{\mathbb{R}^{n-k}} \frac{1}{(a^2 + |y|^2)^{n-1}} dy = \frac{1}{a^{n+k-2}} \int_{\mathbb{R}^{n-k}} \frac{1}{(1 + |y|^2)^{n-1}} dy$$

$$= \frac{\sigma_{n-k}}{2a^{n+k-2}} B(\frac{n-k}{2}, \frac{n+k}{2} - 1),$$

where σ_{n-k} is the volume of the unit $n - k$ dimensional sphere and $B(.,.)$ is the beta function. On the other hand, after a simple computation we find

$$\int_{\mathbb{R}^k} \frac{1}{|x|(|x| + p)^{n+k-2}} dx = \frac{\sigma_k}{p^{n+k+1}} \int_0^\infty \frac{r^{k-2}}{(r+1)^{n+k-2}} dr$$

$$= \frac{\sigma_k}{p^{n+k+1}} B(k-1, n-1).$$

A substitution in (3.104) come to

$$K^{\frac{2(n-1)^2}{n-2}}$$

$$= \left(\frac{n-2}{2}\right)^{2(n-1)} \frac{\sigma_{n-k}}{2} B(\frac{n-k}{2}, \frac{n+k}{2} - 1) \frac{\sigma_k}{p^{n+k+1}} B(k-1, n+k-1)$$

$$= 2^{2k+3}(n-2)^{n-k-3}(k-1)^{n+k+1} \sigma_{n-k} \sigma_k B(\frac{n-k}{2}, \frac{n+k}{2} - 1) B(k-1, n-1).$$
$$(3.105)$$

The proof of Theorem 3.6.2 is complete taking into account the allowed translations in the y variable.

In the above proof we used the following simple ODE lemma, which can be proved by integrating the equation.

Lemma 3.6.5. *Suppose f is a smooth function on $\mathbb{R} \setminus \{0\}$, which is also locally Lipschitz on $\mathbb{R}$, i.e., on any compact interval there is a constant L, such that, $|f(t') - f(t'')| \leq L|t' - t''|$ for any two points t', t'' on this interval. If f satisfies the equation*

$$f'' + \frac{k}{t}f' = \frac{a}{t} + b, \qquad t > 0,$$

where k is a constant $k > 1$ and a, b are L^∞_{loc} functions, then f has bounded first and second order derivatives near the origin.

PART II
Geometry

Chapter 4

Quaternionic contact manifolds - Connection, curvature and qc-Einstein structures

4.1 Introduction

Quaternionic contact structures (qc structures for short), [24; 25], are central for the second part of the book. Such a structure appears naturally as the conformal boundary at infinity of the quaternionic hyperbolic space, see also [136; 83; 64]. Given a qc structure there is a distinguished linear connection [24], called the Biquard connection, which plays a role similar to the Tanaka-Webster connection [169] and [160] in the CR case.

The goal of this Chapter is to define, give examples, and prove some fundamental properties of manifolds equipped with a qc structure.

In Section 4.2 we give a detailed derivation of the construction and properties of the Biquard connection studying its torsion. We describe in detail the structure of the horizontal torsion endomorphism of the Biquard connection showing that this is the obstruction a qc structure to be locally isomorphic to (positive or negative) 3-Sasakian structure. We derive local structure equations of a qc-manifold using the basic objects of the structure - the contact forms, the corresponding Reeb vector fields and the horizontal torsion endomorphism. In Section 4.3 we continue with a study of the corresponding qc-Ricci tensor and qc-scalar curvature, defined in (4.63). With the help of the Bianchi's identities we derive fundamental properties and formulas for the qc curvature, qc-Ricci tensor and the connection 1-forms. In particular, we shall see that the two components of the horizontal torsion endomorphism and the qc-scalar curvature determine completely the curvature of the Biquard connection in vertical directions as well as the traceless part of the qc-Ricci tensor. In Section 4.3.4 we introduce the quaternionic Heisenberg group as the flat model of a qc structure in addition to some other noteworthy examples.

In the subsequent Section 4.4 we present an in-depth study of qc-Einstein structures, introduced in [94]. These special qc structures turn out to be not only of crucial importance when studying the Yamabe equation, but also a direct link to the extensively studied 3-Sasakian structures. We shall find the local structure equations of a qc-Einstein manifold and show the (local) equivalence with a 3-Sasakian manifold by considering cones over qc structures, Theorems 4.4.3 and 4.4.4.

Convention 4.1.1. Everywhere in Part 2 of the book we use the following conventions:

a) X, Y, Z, U denote horizontal vector fields, $X, Y, Z, U \in H$;
b) A, B, C, D stand for arbitrary vector fields, $A, B, C, D \in \Gamma(TM)$;
c) $\{e_1, \dots, e_{4n}\}$ denotes an orthonormal basis of the horizontal space H;
d) The summation convention over repeated vectors from the basis $\{e_1, \dots, e_{4n}\}$ will be used. For example, for a (0,4)-tensor P, the formula $k = P(e_b, e_a, e_a, e_b)$ means $k = \sum_{a,b=1}^{4n} P(e_b, e_a, e_a, e_b)$;
e) The triple (i, j, k) denotes any cyclic permutation of $(1, 2, 3)$. In particular, any equation involving i, j, k holds for any such permutation.
f) s and t will be any numbers from the set $\{1, 2, 3\}$, $\quad s, t \in \{1, 2, 3\}$.

4.2 Quaternionic contact structures and the Biquard connection

The notion of *Quaternionic Contact Structure* has been introduced by O. Biquard in [24] and [25]. Namely, a quaternionic contact structure (qc structure for short) on a $(4n + 3)$-dimensional smooth manifold M is a codimension 3 distribution H, such that, at each point $p \in M$ the nilpotent step two Lie algebra $H_p \oplus (T_pM/H_p)$ is isomorphic to the quaternionic Heisenberg algebra $\mathbb{H}^n \oplus Im\ \mathbb{H}$. The nilpotent Lie algebra structures on $H_p \oplus (T_pM/H_p)$ is defined by

$$[V_1, V_2] = \begin{cases} \pi_{T_pM/H_p}[\widetilde{V_1}, \widetilde{V_2}], & \text{if } V_1,\ V_1 \in H_p \\ 0, & \text{otherwise,} \end{cases}$$

where $\widetilde{V_1}, \widetilde{V_2}$ are two vector fields, such that, $\widetilde{V_j}(p) = V_j$, $j = 1, 2$. The quaternionic Heisenberg algebra structure on $\mathbb{H}^n \oplus Im\ \mathbb{H}$ is obtained by the identification of $\mathbb{H}^n \oplus Im\ \mathbb{H}$ with the algebra of the left invariant vector fields on the quaternionic Heisenberg group, see Section 4.3.4. In particular, the Lie bracket is given by the formula $[(q_o, \omega_o), (q, \omega)] = 2\ Im\ q_o \cdot \bar{q}$, where

$q = (q^1, q^2, \ldots, q^n)$, $q_o = (q_o^1, q_o^2, \ldots, q_o^n) \in \mathbb{H}^n$ and ω, $\omega_o \in Im \ \mathbb{H}$ with $q_o \cdot \bar{q} = \sum_{\alpha=1}^{n} q_o^\alpha \cdot \overline{q^\alpha}$, see Section 4.3.4 for notations concerning the qc group $G(\mathbb{H})$. It is important to observe that if M has a quaternionic contact structure as above then the definition implies that the distribution H and its commutators generate the tangent space at every point. A manifold M with a structure as above will be called also quaternionic contact manifold (qc manifold) and denoted by $(M, [g], \mathbb{Q})$.

The following is another, more explicit, definition of a quaternionic contact structure.

Definition 4.2.1. [24] A quaternionic contact (qc) manifold $(M, [g], \mathbb{Q})$ is a $4n + 3$-dimensional manifold M with a codimension three distribution H such that

   i) H has an conformal $Sp(n)Sp(1)$ structure, that is, it is equipped with a conformal class of Riemannian metrics $[g]$ and a rank-three bundle $\mathbb{Q}$ consisting of (1,1)-tensors on H locally generated by three almost complex structures I_1, I_2, I_3 on H satisfying the identities of the imaginary unit quaternions, $I_1 I_2 = -I_2 I_1 = I_3$, $I_1 I_2 I_3 = -id_{|_H}$ which are hermitian compatible with any metric $g \in [g]$, $g(I_s \cdot, I_s \cdot) = g(\cdot, \cdot)$.

   ii) H is locally given as the kernel of a 1-form $\eta = (\eta_1, \eta_2, \eta_3)$ with values in $\mathbb{R}^3$ and the following compatibility condition holds

$$2g(I_s X, Y) = d\eta_s(X, Y), \quad g \in [g]. \tag{4.1}$$

The fundamental 2-forms ω_s are defined by

$$2\omega_{s|H} = d\eta_{s|H}, \quad \xi \lrcorner \omega_s = 0, \quad \xi \in V, \tag{4.2}$$

where V is defined in (4.7). Given a qc structure we have a 2-sphere bundle Q over M of almost complex structures on H, such that, locally $Q = \{aI_1 + bI_2 + cI_3 : a^2 + b^2 + c^2 = 1\}$.

If in some local chart $\bar{\eta}$ is another form, with corresponding $\bar{g} \in [g]$ and almost complex structures $\bar{I}_s$, then $\bar{\eta} = \mu \Psi \eta$ for some $\Psi \in SO(3)$ and a positive function μ. Typical examples of manifolds with qc structures are totally umbilical hypersurfaces in quaternionic Kähler or hyperkähler manifold.

It is instructive to consider the case when there is a globally defined one-form η. The obstruction to the global existence of η is encoded in the first Pontrjagin class [4]. Besides clarifying the notion of a qc manifold, most of the time, for example when considering the Yamabe equation, we shall work with a qc structure for which we have a fixed globally defined contact form. In this case, if we rotate the $\mathbb{R}^3$-valued contact form and the almost complex structures by the same rotation we obtain again a contact form, almost complex structures and a metric on H (the latter is unchanged) satisfying the above conditions. On the other hand, it is important to observe that given a contact form the almost complex structures and the horizontal metric are unique if they exist. Finally, if we are given the horizontal bundle and a metric on it, there exists at most one sphere of associated contact forms with a corresponding sphere Q of almost complex structures [24]. This is the content of the next Lemma.

Lemma 4.2.2. *[24] Let $(M, [g], \mathbb{Q})$ be a qc manifold. Then:*

a) *If (η, I_s, g) and (η, I_s', g') are two qc structures on M, then $I_s = I_s'$ and $g = g'$.*

b) *If (η, g) and (η', g) are two qc structures on M with $Ker(\eta) = Ker(\eta') = H$ then $Q = Q'$ and $\eta' = \Psi\eta$ for some matrix $\Psi \in SO(3)$ with smooth functions as entries.*

Proof. Let $g, d\eta_{1|H}, d\eta_{2|H}, d\eta_{3|H}, I_1, I_2, I_3$ be given by the matrices $G, N_1, N_2, N_3, J_1, J_2, J_3$, respectively. From (4.1) it follows $2GJ_s = N_s$ and $J_k = J_iJ_j = -J_i^{-1}G^{-1}GJ_j = -(GJ_i)^{-1}(GJ_j) = -N_i^{-1}N_j$, which proofs a).

The condition $Ker(\eta) = Ker(\eta') = H$ implies that $\eta_s' = \sum_{t=1}^3 \Psi_{st}\eta_t$ for some matrix $\Psi_{st} \in GL(3)$. Applying the exterior derivative, we find $d\eta_s' = \sum_{t=1}^3 (d\Psi_{st} \wedge \eta_t + \Psi_{st}\,d\eta_t)$ which restricted to H gives $g(I_sX,Y) = \sum_{t=1}^3 \Psi_{st}g(I_tX,Y)$. Equivalently, $I_s = \sum_{t=1}^3 \Psi_{st}I_t$. Hence, $\Psi_{st} \in SO(3)$. $\square$

Besides the non-uniqueness due to the action of $SO(3)$, the 1-form η can be changed by a conformal factor, in the sense that if η is a form for which we can find associated almost complex structures and metric g as above, then for any $\Psi \in SO(3)$ and a positive function μ, the form $\mu\Psi\eta$ also has an associated complex structures and metric. In particular, when $\mu = 1$ we obtain a whole unit sphere of contact forms, and we shall denote, as already mentioned, by Q the corresponding sphere bundle of associated triples of almost complex structures. With the above consideration in mind we introduce the following notation.

Notation 4.2.3. We shall denote with (M, η) a qc manifold with a fixed globally defined contact form. $(M, g, \mathbb{Q})$ will denote a qc manifold with a fixed metric g and a quaternionic bundle $\mathbb{Q}$ on H. In this case we have in fact a $Sp(n)Sp(1)$ structure on the horizontal distribution H. Correspondingly, we shall denote with η any (locally defined) associated contact form.

We recall the definition of the Lie groups $Sp(n)$, $Sp(1)$ and $Sp(n)Sp(1)$. Let us identify $\mathbb{H}^n = \mathbb{R}^{4n}$ and let $\mathbb{H}$ acts on $\mathbb{H}^n$ by right multiplications, $\lambda(q)(W) = W \cdot q^{-1}$. This defines a homomorphism λ : {unit quaternions} $\longrightarrow SO(4n)$ with the convention that $SO(4n)$ acts on $\mathbb{R}^{4n}$ on the left. The image is the Lie group $Sp(1)$. Let $\lambda(i) = I_0, \lambda(j) = J_0, \lambda(k) = K_0$. The Lie algebra of $Sp(1)$ is $sp(1) = span\{I_0, J_0, K_0\}$. The group $Sp(n)$ is $Sp(n) = \{O \in SO(4n) : OB = BO$ for all $B \in Sp(1)\}$ or $Sp(n) = \{O \in GL(n, \mathbb{H}) : O\bar{O}^t = I\}$, and $O \in Sp(n)$ acts by $(q^1, q^2, \ldots, q^n)^t \mapsto O(q^1, q^2, \ldots, q^n)^t$. Denote by $Sp(n)Sp(1)$ the product of the two groups in $SO(4n)$. Abstractly, $Sp(n)Sp(1) = (Sp(n) \times Sp(1))/\mathbb{Z}_2$. The Lie algebra of the group $Sp(n)Sp(1)$ is $sp(n) \oplus sp(1)$.

Any endomorphism Ψ of H can be decomposed with respect to the quaternionic structure $(\mathbb{Q}, g)$ uniquely into four $Sp(n)$-invariant parts $\Psi = \Psi^{+++} + \Psi^{+--} + \Psi^{-+-} + \Psi^{--+}$, where Ψ^{+++} commutes with all three I_i, Ψ^{+--} commutes with I_1 and anti-commutes with the others two and etc. Explicitly,

$$
\begin{aligned}
4\Psi^{+++} &= \Psi - I_1\Psi I_1 - I_2\Psi I_2 - I_3\Psi I_3, \\
4\Psi^{+--} &= \Psi - I_1\Psi I_1 + I_2\Psi I_2 + I_3\Psi I_3, \\
4\Psi^{-+-} &= \Psi + I_1\Psi I_1 - I_2\Psi I_2 + I_3\Psi I_3, \\
4\Psi^{--+} &= \Psi + I_1\Psi I_1 + I_2\Psi I_2 - I_3\Psi I_3.
\end{aligned}
\tag{4.3}
$$

The two $Sp(n)Sp(1)$-invariant components are given by

$$
\Psi_{[3]} = \Psi^{+++}, \quad \Psi_{[-1]} = \Psi^{+--} + \Psi^{-+-} + \Psi^{--+}
$$

with the following characterising conditions

$$
\begin{aligned}
\Psi = \Psi_{[3]} &\iff 3\Psi + I_1\Psi I_1 + I_2\Psi I_2 + I_3\Psi I_3 = 0, \\
\Psi = \Psi_{[-1]} &\iff \Psi - I_1\Psi I_1 - I_2\Psi I_2 - I_3\Psi I_3 = 0.
\end{aligned}
\tag{4.4}
$$

Denoting the corresponding $(0,2)$ tensor via g by the same letter one sees that the $Sp(n)Sp(1)$-invariant components are the projections on the eigenspaces of the Casimir operator

$$
\dagger = I_1 \otimes I_1 + I_2 \otimes I_2 + I_3 \otimes I_3
\tag{4.5}
$$

corresponding, respectively, to the eigenvalues 3 and -1, see [39]. If $n = 1$ then the space of symmetric endomorphisms commuting with all I_s is 1-dimensional, i.e. the [3]-component of any symmetric endomorphism Ψ on

H is proportional to the identity, $\Psi_3 = \frac{|\Psi|^2}{4} Id_{|H}$. Note here that each of the three 2-forms ω_s belongs to its $[-1]$-component, $\omega_s = \omega_{s[-1]}$ and constitute a basis of the lie algebra $sp(1)$.

Consider the orthogonal complement $(sp(n) \oplus sp(1))^{\perp} \subset so(4n)$ of the lie algebra $(sp(n) \oplus sp(1)) \subset so(4n)$ with respect to the standard inner product $<, >$ on $so(4n)$ coming from the standard inner product in the lie algebra $gl(4n)$ of the general linear group defined by $< A, B > = tr(B^*A) = < A(e_a), B(e_a) >$, $A, B \in gl(4n)$. It is known that a skew-symmetric endomorphism $A \in so(4n)$ considered as an element of the orthogonal lie algebra $so(4n)$ belongs to the orthogonal complement $(sp(n) \oplus sp(1))^{\perp} \subset so(4n)$ if and only if A coincides with the completely trace-free part of its $[-1]$-component. More precisely, we have
$$A \in (sp(n) \oplus sp(1))^{\perp} \iff A = A_{[-1]} - A_{sp(1)}, \qquad (4.6)$$
where $A_{sp(1)}$ denotes the orthogonal projection of A onto $sp(1)$ given by $4nA_{sp(1)} = \sum_{s=1}^{3} A(e_a, I_s e_a)\omega_s$.

There exists a canonical connection compatible with a given quaternionic contact structure. This connection was discovered by O. Biquard [24] when the dimension $(4n + 3) > 7$ and by D. Duchemin [61] in the 7-dimensional case. The next result due to O. Biquard is crucial in the quaternionic contact geometry.

Theorem 4.2.4. *[24] Let $(M, g, \mathbb{Q})$ be a quaternionic contact manifold of dimension $4n + 3 > 7$ and a fixed metric g on H in the conformal class $[g]$. Then there exists a unique connection ∇ with torsion T on M^{4n+3} and a unique supplementary subspace V to H in TM, such that:*

   i) *∇ preserves the decomposition $H \oplus V$ and the metric g;*
   ii) *for $X, Y \in H$, one has $T(X, Y) = -[X, Y]_{|V}$;*
   iii) *∇ preserves the $Sp(n)Sp(1)$-structure on H, i.e., $\nabla g = 0$ and $\nabla \sigma \in \Gamma(\mathbb{Q})$ for a section $\sigma \in \Gamma(\mathbb{Q})$;*
   iv) *for $\xi \in V$, the torsion endomorphism $T(\xi, .)_{|H}$ of H lies in $(sp(n) \oplus sp(1))^{\perp} \subset gl(4n)$;*
   v) *the connection on V is induced by the natural identification φ of V with the subspace $sp(1)$ of the endomorphisms of H, i.e. $\nabla \varphi = 0$.*

We shall call the above connection *the Biquard connection*. Biquard [24] also described the supplementary subspace V explicitly, namely, locally V is generated by vector fields $\{\xi_1, \xi_2, \xi_3\}$, such that
$$\eta_s(\xi_t) = \delta_{st}, \qquad (\xi_s \lrcorner d\eta_s)_{|H} = 0,$$
$$(\xi_s \lrcorner d\eta_t)_{|H} = -(\xi_t \lrcorner d\eta_s)_{|H}. \qquad (4.7)$$

The vector fields ξ_1, ξ_2, ξ_3 are called Reeb vector fields or fundamental vector fields.

If the dimension of M is seven, the conditions (4.7) do not always hold. Duchemin shows in [61] that if we assume, in addition, the existence of Reeb vector fields as in (4.7), then Theorem 4.2.4 holds. Henceforth, by a quaternionic contact (qc) structure in dimension 7 we shall always mean a qc structure satisfying (4.7).

Notice that equations (4.7) are invariant under the natural $SO(3)$ action. Using the Reeb vector fields we extend g to a metric on M by requiring

$$span\{\xi_1, \xi_2, \xi_3\} = V \perp H \text{ and } g(\xi_s, \xi_k) = \delta_{sk}. \qquad (4.8)$$

The extended metric does not depend on the action of $SO(3)$ on V, but it changes in an obvious manner if η is multiplied by a conformal factor. Clearly, the Biquard connection preserves the extended metric on TM, $\nabla g = 0$. We shall also extend the quternionic structure by setting $I_{s|V} = 0$.

We state and prove the Biquard theorem in a slightly different way. First recall that the torsion tensor T of a linear connection ∇ is defined by

$$T(A, B) = \nabla_A B - \nabla_B A - [A, B].$$

We denote the torsion tensor of type (3,0) by the same letter,

$$T(A, B, C) = g(T(A, B), C) = g(\nabla_A B - \nabla_B A - [A, B], C). \qquad (4.9)$$

Theorem 4.2.5. *Let $(M, g, \mathbb{Q})$ be a quaternionic contact manifold of dimension $4n + 3$ and a fixed metric g on H in the conformal class $[g]$. Suppose the Reeb vector fields do exists also if the dimension is seven and denote with $V = span\{\xi_1, \xi_2, \xi_3\}$ the vertical space to H. Then there exists a unique connection ∇ on M with torsion T preserving the extendet metric g, $\nabla g = 0$ satisfying the following conditions*

$$\nabla I_i = -\alpha_j \otimes I_k + \alpha_k \otimes I_j; \quad \nabla \xi_i = -\alpha_j \otimes \xi_k + \alpha_k \otimes \xi_j; \qquad (4.10)$$

$$T(X, Y) = -[X, Y]_{|V} = 2 \sum_{s=1}^{3} \omega_s(X, Y)\xi_s; \qquad (4.11)$$

$$T(\xi, X, \xi_s) = 0 \quad \Longleftrightarrow \quad T(\xi, X) = \nabla_\xi X - [\xi, X]_{|H}, \quad \xi \in V, \qquad (4.12)$$

where α_s are 1-forms on M.
The skew-symmetric part $S(\xi, X, Y)$ of $T(\xi, X, Y)$ on H lies in $(sp(n) \oplus sp(1))^\perp \subset so(4n)$,

$$S(\xi, X, Y) = \frac{1}{2}(T(\xi, X, Y) - T(\xi, Y, X)) \in (sp(n) \oplus sp(1))^\perp. \qquad (4.13)$$

Proof. Since the connection ∇ preserves the extended metric it is sufficient to determine the torsion in terms of the data supplied by the quaternionic contact structure and the Reeb vector fields. The difference between the Levi-Civita connection ∇^g of the extended metric g and ∇ is given by the well known formula

$$g(\nabla_A B, C) - g(\nabla^g_A B, C) =$$
$$\frac{1}{2}\Big(T(A, B, C) - T(B, C, A) + T(C, A, B)\Big). \quad (4.14)$$

The equations in (4.10) imply that ∇ preserves the distribution H. The latter follows from the identities

$$0 = Ag(\xi_s, X) = g(\nabla_A \xi_s, X) + g(\xi_s, \nabla_A X) = g(\xi_s, \nabla_A X).$$

From (4.10) we calculate

$$d\eta_i(A, B) = (-\alpha_j \wedge \eta_k + \alpha_k \wedge \eta_j)(A, B) + T(A, B, \xi_i). \quad (4.15)$$

Set $A = \xi_j, B = Y$ into (4.15) and apply (4.12) to get (see also [24])

$$d\eta_i(\xi_j, Y) = -\alpha_k(Y) = -d\eta_j(\xi_i, Y). \quad (4.16)$$

Similarly, we obtain from (4.15) the following identities

$$d\eta_i(\xi_i, \xi_j) = \alpha_k(\xi_i) + T(\xi_i, \xi_j, \xi_i),$$
$$d\eta_i(\xi_i, \xi_k) = -\alpha_j(\xi_i) + T(\xi_i, \xi_k, \xi_i), \quad (4.17)$$

$$d\eta_i(\xi_j, \xi_k) = -\alpha_j(\xi_j) - \alpha_k(\xi_k) + T(\xi_j, \xi_k, \xi_i). \quad (4.18)$$

Take a cyclic permutation of i, j, k in (4.18), summing the second and the third and subtracting the first, we have

$$2\alpha_i(\xi_i) = d\eta_i(\xi_j, \xi_k) - d\eta_j(\xi_k, \xi_i) - d\eta_k(\xi_i, \xi_j)$$
$$+ T(\xi_i, \xi_j, \xi_k) + T(\xi_k, \xi_i, \xi_j) - T(\xi_j, \xi_k, \xi_i). \quad (4.19)$$

Decomposing the torsion endomorphism into symmetric part T^0 and skew-symmetric part S we write

$$T(\xi_s, X, Y) = T^0(\xi_s, X, Y) + S(\xi_s, X, Y), \quad (4.20)$$

where $T^0(\xi_s, X, Y) = \frac{1}{2}(T(\xi_s, X, Y) + T(\xi_s, Y, X))$ and S is determined with (4.13).

The Lie derivatives $\mathcal{L}_\xi g, \mathcal{L}_{\xi_t}\omega_s, \mathcal{L}_{\xi_s}I_t$ can be expressed in terms of the torsion. Applying (4.1), (4.10) and (4.9), we calculate

$$
\begin{aligned}
g((\mathcal{L}_{\xi_s}I_i)X,Y) &= g([\xi_s,I_iX],Y) - g(I_i[\xi_s,X],Y) \\
&= g(\nabla_{\xi_s}I_iX - \nabla_{I_iX}\xi_s - T(\xi_s,I_iX),Y) \\
&\quad - g(I_i\nabla_{\xi_s}X - I_i\nabla_X\xi_s - I_iT(\xi_s,X),Y) \\
&= g((\nabla_{\xi_s}I_i)X,Y) - T(\xi_s,I_iX,Y) - T(\xi_s,X,I_iY) = -\alpha_j(\xi_s)\omega_k(X,Y) \\
&\quad + \alpha_k(\xi_s)\omega_j(X,Y) - T(\xi_s,I_iX,Y) - T(\xi_s,X,I_iY). \quad (4.21)
\end{aligned}
$$

In a similar way, we get

$$
\begin{aligned}
(\mathcal{L}_{\xi_j}\omega_i)(X,Y) &= \xi_j g(I_iX,Y) - g(I_i[\xi_j,X],Y) - g(I_iX,[\xi_j,Y]) \\
&= g((\nabla_{\xi_j}I_i)X,Y) - T(\xi_j,X,I_iY) + T(\xi_j,Y,I_iX) = -\alpha_j(\xi_j)\omega_k(X,Y) \\
&\quad + \alpha_k(\xi_j)\omega_j(X,Y) - T(\xi_j,X,I_iY) + T(\xi_j,Y,I_iX). \quad (4.22)
\end{aligned}
$$

$$
\begin{aligned}
(\mathcal{L}_{\xi_i}\omega_j)(X,Y) &= -\alpha_k(\xi_i)\omega_i(X,Y) + \alpha_i(\xi_i)\omega_k(X,Y) \\
&\quad - T(\xi_i,X,I_jY) + T(\xi_i,Y,I_jX), \quad (4.23)
\end{aligned}
$$

$$
\begin{aligned}
(\mathcal{L}_{\xi_i}\omega_i)(X,Y) &= -\alpha_j(\xi_i)\omega_k(X,Y) + \alpha_k(\xi_i)\omega_j(X,Y) \\
&\quad - T(\xi_i,X,I_iY) + T(\xi_i,Y,I_iX). \quad (4.24)
\end{aligned}
$$

$$
(\mathcal{L}_{\xi_s}g)(X,Y) = T(\xi_s,X,Y) + T(\xi_s,Y,X) = 2T^0(\xi_s,X,Y). \quad (4.25)
$$

The symmetric part of the torsion endomorphism is determined entirely by the Lie derivative of the metric according to (4.25). The skew-symmetric part S is trace-free because of (4.13) and (4.6). Using the identity

$$
g((\mathcal{L}_{\xi_s}I_i)X,Y) = -g((\mathcal{L}_{\xi_s}I_i)I_iX,I_iY) \quad (4.26)
$$

we conclude from (4.21) and (4.25) that $T^0(\xi_s,.,.)$ is completely trace-free since the Lie derivative commutes with taking the trace, $(\mathcal{L}_{\xi_s}g)(e_a,e_a) = \mathcal{L}_{\xi_s}4n = 0$ [112]. Hence, we have [24]

$$
T(\xi_s,e_a,a_a) = T(\xi_s,e_a,I_te_a) = 0. \quad (4.27)
$$

On the other hand, the Cartan's formula yields

$$
\mathcal{L}_{\xi_k}\omega_l = \xi_k\lrcorner(d\omega_l) + d(\xi_k\lrcorner\omega_l). \quad (4.28)
$$

A direct calculation using (4.2) gives

$$
2\omega_l = (d\eta_l)_{|H} = d\eta_l - \sum_{s=1}^{3}\eta_s \wedge (\xi_s\lrcorner d\eta_l) + \sum_{s<t}d\eta_l(\xi_s,\xi_t)\eta_s \wedge \eta_t. \quad (4.29)
$$

Combining (4.29) and (4.28), we obtain after a short calculation using (4.7) and (4.16) that

$$(\mathcal{L}_{\xi_i}\omega_i)(X,Y) = d\eta_i(\xi_i,\xi_j)\omega_j(X,Y) + d\eta_i(\xi_i,\xi_k)\omega_k(X,Y) \tag{4.30}$$

and

$$
\begin{aligned}
2(\mathcal{L}_{\xi_i}\omega_j)(X,Y) &= (d(\xi_i\lrcorner d\eta_j) - (\xi_i\lrcorner d\eta_k)\wedge(\xi_k\lrcorner d\eta_j))(X,Y) \\
&= (d\alpha_k + \alpha_i \wedge \alpha_j)(X,Y) - d\eta_j(\xi_i,[X,Y]_{|V}) + \alpha_k([X,Y]_{|V}) \\
&= (d\alpha_k + \alpha_i \wedge \alpha_j)(X,Y) + 2\omega_j(X,Y)d\eta_j(\xi_i,\xi_j) + 2\omega_k(X,Y)d\eta_j(\xi_i,\xi_k) \\
&\quad - 2\omega_i(X,Y)\alpha_k(\xi_i) - 2\omega_j(X,Y)\alpha_k(\xi_j) - 2\omega_k(X,Y)\alpha_k(\xi_k) \tag{4.31}
\end{aligned}
$$

and also

$$
\begin{aligned}
2(\mathcal{L}_{\xi_j}\omega_i)(X,Y) &= (d(\xi_j\lrcorner d\eta_i) - (\xi_j\lrcorner d\eta_k)\wedge(\xi_k\lrcorner d\eta_i))(X,Y) \\
&= -(d\alpha_k + \alpha_i \wedge \alpha_j)(X,Y) + 2\omega_i(X,Y)d\eta_i(\xi_j,\xi_i) + 2\omega_k(X,Y)d\eta_i(\xi_j,\xi_k) \\
&\quad + 2\omega_i(X,Y)\alpha_k(\xi_i) + 2\omega_j(X,Y)\alpha_k(\xi_j) + 2\omega_k(X,Y)\alpha_k(\xi_k). \tag{4.32}
\end{aligned}
$$

From (4.24) and (4.30) together with (4.17) we have

$$
\begin{aligned}
T(\xi_i,\xi_k,\xi_i)\omega_k(X,Y) &+ T(\xi_i,\xi_j,\xi_i)\omega_j(X,Y) \\
&- T(\xi_i,X,I_iY) + T(\xi_i,Y,I_iX) = 0. \tag{4.33}
\end{aligned}
$$

Taking suitable traces of (4.33) and using (4.13) we conclude

$$T(\xi_i,\xi_k,\xi_i) = T(\xi_i,\xi_j,\xi_i) = 0, \tag{4.34}$$

$$T(\xi_i,X,I_iY) = T(\xi_i,Y,I_iX). \tag{4.35}$$

Substitute (4.34) into (4.17) to get

$$d\eta_i(\xi_i,\xi_j) = \alpha_k(\xi_i), \qquad d\eta_i(\xi_i,\xi_k) = -\alpha_j(\xi_i). \tag{4.36}$$

When we add (4.31) and (4.32) and then compare the result with the sum of (4.22) and (4.23) we see

$$
\begin{aligned}
(\mathcal{L}_{\xi_i}\omega_j &+ \mathcal{L}_{\xi_j}\omega_i)(X,Y) \\
&= d\eta_i(\xi_j,\xi_i)\omega_i(X,Y) + d\eta_j(\xi_i,\xi_j)\omega_j(X,Y) \\
&\quad + (d\eta_i(\xi_j,\xi_k) + d\eta_j(\xi_i,\xi_k))\omega_k(X,Y) \\
&= (\alpha_i(\xi_i) - \alpha_j(\xi_j))\omega_k(X,Y) + \alpha_k(\xi_j)\omega_j(X,Y) - \alpha_k(\xi_i)\omega_i(X,Y) \\
&\quad - T(\xi_j,X,I_iY) + T(\xi_j,Y,I_iX) - T(\xi_i,X,I_jY) + T(\xi_i,Y,I_jX). \tag{4.37}
\end{aligned}
$$

Applying (4.36) to (4.37) yields

$$
\begin{aligned}
\Big[d\eta_i(\xi_j,\xi_k) &+ d\eta_j(\xi_i,\xi_k) - \alpha_i(\xi_i) + \alpha_j(\xi_j)\Big]\omega_k(X,Y) \\
&= -T(\xi_j,X,I_iY) + T(\xi_j,Y,I_iX) - T(\xi_i,X,I_jY) + T(\xi_i,Y,I_jX). \tag{4.38}
\end{aligned}
$$

Considering the traceless part of (4.38) together with (4.27) shows that

$$T(\xi_j, X, I_i Y) - T(\xi_j, Y, I_i X) + T(\xi_i, X, I_j Y) - T(\xi_i, Y, I_j X) = 0, \quad (4.39)$$

while the trace part of (4.38) gives

$$\alpha_i(\xi_i) - \alpha_j(\xi_j) = d\eta_i(\xi_j, \xi_k) - d\eta_j(\xi_k, \xi_i). \quad (4.40)$$

A substitution of (4.19) in (4.40) yields

$$T(\xi_j, \xi_k, \xi_i) = T(\xi_k, \xi_i, \xi_j) = -\lambda. \quad (4.41)$$

Consequently, (4.41) together with (4.40) implies

$$2\alpha_i(\xi_i) = d\eta_i(\xi_j, \xi_k) - d\eta_j(\xi_k, \xi_i) - d\eta_k(\xi_i, \xi_j) - \lambda. \quad (4.42)$$

The properties of the torsion endomorphism are encoded into (4.35) and (4.39). Decomposing (4.35) into symmetric and skew-symmetric parts gives

$$\begin{aligned} T^0(\xi_s, I_s X, I_s Y) &= -T^0(\xi_s, X, Y); \\ S(\xi_s, I_s X, I_s Y) &= S(\xi_s, X, Y). \end{aligned} \quad (4.43)$$

By considering the symmetric and skew-symmetric parts of (4.39), taking also into account (4.43), we obtain the next two identities:

$$T^0(\xi_i, I_j X, I_j Y) + T^0(\xi_i, Y, X) - \\ T^0(\xi_j, X, I_k Y) - T^0(\xi_j, Y, I_k X) = 0, \quad (4.44)$$

$$S(\xi_i, I_j X, I_j Y) + S(\xi_i, Y, X) + \\ S(\xi_j, X, I_k Y) - S(\xi_j, Y, I_k X) = 0. \quad (4.45)$$

Let $T_s^0(X, Y) \doteq T^0(\xi_s, I_s X, Y)$ and consider the following tensors defined on the horizontal space H by the formulas

$$\begin{aligned} T^0(X, Y) &= T_i^0(X, Y) + T_j^0(X, Y) + T_k^0(X, Y), \\ U_s(X, Y) &= -S(\xi_s, I_s X, Y). \end{aligned} \quad (4.46)$$

The tensor T^0 does not depend on the particular choice of the Reeb vector fields and is invariant under the natural action of $SO(3)$. Indeed, if $\bar\eta_s = \sum_{t=1}^3 \Psi_{st}\eta_t$, $\Psi_{st} \in SO(3)$, we have $\bar\xi_s = \sum_{t=1}^3 \Psi_{st}\xi_t$ and $\bar I_s = \sum_{t=1}^3 \Psi_{st}I_t$ which substituted into the first equality in (4.46) does not change it.

The first equality of (4.43) yields

$$T_s^0(X, Y) = T^0(\xi_s, I_s X, Y) = T^0(\xi_s, X, I_s Y) = T_s^0(Y, X),$$
$$T_s^0(I_s X, I_s Y) = -T^0(\xi_s, X, I_s Y) = -T^0(\xi_s, I_s X, Y) = -T_s^0(X, Y),$$
$$(4.47)$$

which tells us that the tensor T_s^0 is symmetric and anti-invariant with respect to I_s.

The second equality of (4.43) gives

$$U_s(X,Y) = -S(\xi_s, I_s X, Y) = S(\xi_s, X, I_s Y) = U_s(Y, X),$$
$$U_s(I_s X, I_s Y) = S(\xi_s, X, I_s Y) = -S(\xi_s, I_s X, Y) = U_s(X, Y), \qquad (4.48)$$

which shows that the tensor U_s is symmetric and invariant with respect to I_s.

Using again (4.43), (4.47) and (4.48) we rewrite (4.44) and (4.45) as follows

$$T_i^0(X,Y) - T_i^0(I_j X, I_j Y) = T_j^0(X,Y) + T_j^0(I_k X, I_k Y),$$
$$U_i(X,Y) + U_i(I_j X, I_j Y) = U_j(X,Y) + U_j(I_k X, I_k Y). \qquad (4.49)$$

First we describe the properties of the symmetric tensor T^0. We have

Lemma 4.2.6. *The symmetric $SO(3)$-invariant tensor T^0 is completely trace-free, satisfies the following identity*

$$T^0(X,Y) + T^0(I_i X, I_i Y) + T^0(I_j X, I_j Y) + T^0(I_k X, I_k Y) = 0 \qquad (4.50)$$

and determines the symmetric part of the torsion endomorphism via the formula

$$T^0(\xi_s, I_s X, Y) = \frac{1}{4}\Big(T^0(X,Y) - T^0(I_s X, I_s Y)\Big). \qquad (4.51)$$

Proof of Lemma 4.2.6. The first part follows from (4.27). Using the first equality in (4.43) we rewrite the first equality in (4.49) in the form

$$-T_i^0(X,Y) - T_i^0(I_k X, I_k Y) = T_j^0(I_i X, I_i Y) - T_j^0(X,Y).$$

The cyclic sum of $\{i, j, k\}$ in the latter equality as well as in the first equality in (4.49) gives

$$T_i^0(I_k X, I_k Y) + T_j^0(I_i X, I_i Y) + T_k^0(I_j X, I_j Y) = 0,$$
$$T_i^0(I_j X, I_j Y) + T_j^0(I_k X, I_k Y) + T_k^0(I_i X, I_i Y) = 0.$$

Summing up the above two equalities using the first equality in (4.43) gives (4.50) according to the definition of T^0 given in the first equality of (4.46) which proves the first part of the lemma.

The second part of the lemma follows from the (4.46), (4.47) and the just proved (4.51) in a straightforward way. The lemma is proved. $\square$

Lemma 4.2.7.

a) *The three symmetric tensors U_s are all equal, $U_s = U_t$.*

b) *The symmetric trace-free tensor defined by $U = U_i$ commutes with each of the almost complex strictures I_s,*

$$U(X,Y) = U(I_iX, I_iY) = U(I_jX, I_jY) = U(I_kX, I_kY) \qquad (4.52)$$

and therefore it is $SO(3)$-invariant.

c) *If the dimension of M is seven then $U = 0$.*

d) *The $SO(3)$-invariant tensor U determines the skew-symmetric part of the torsion by*

$$S(\xi_s, X, Y) = U(I_sX, Y). \qquad (4.53)$$

Proof. Since $S(\xi_s, ., .) \in (sp(n) \oplus sp(1))^\perp \subset so(4n)$, we conclude applying (4.6) that S is completely trace-free and $S = S_{[-1]}$, i.e. it satisfies the equality

$$S(\xi_s, X, Y) + S(\xi_s, I_iX, I_iY) + S(\xi_s, I_jX, I_jY) + S(\xi_s, I_kX, I_kY) = 0.$$

Take $\xi_s = \xi_1$ in the above equality and write the result in terms of U_i according to (4.46) to get

$$U_i(I_iX, Y) - U_i(X, I_iY) + U_i(I_kX, I_jY) - U_i(I_jX, I_kY) = 0.$$

An application of (4.48) yields $U_i(X, Y) = U_i(I_iX, I_iY) = U_i(I_jX, I_jY)$ which combined with (4.49) proofs (4.52).

The vanishing of U in dimension seven follows from the fact that in this case H is four dimensional and any trace-free symmetric tensor commuting with all I_s on a four dimensional vector space must vanish.

Recalling (4.46) completes the proof of Lemma 4.2.7. $\qquad \square$

Now we express the tensor U in terms of Lie derivatives of the structure thus determining the skew-symmetric part of the torsion endomorphism. With the help of Lemma 4.2.6 and Lemma 4.2.7 write (4.21) in terms of T^0 and U to get

$$g((\mathcal{L}_{\xi_j} I_i)I_kX, Y) = \alpha_j(\xi_j)g(X,Y) + \alpha_k(\xi_j)\omega_i(X,Y)$$
$$- 2U(X,Y) + \frac{1}{2}\Big(T^0(X,Y) + T^0(I_iX, I_iY)\Big). \qquad (4.54)$$

The equation (4.54) together with an application of Lemma 4.2.6 and Lemma 4.2.7 yields

$$g((\mathcal{L}_{\xi_j} I_i)I_kX, Y) - g((\mathcal{L}_{\xi_j} I_i)X, I_kY) =$$
$$2\alpha_j(\xi_j)g(X,Y) - 4U(X,Y). \qquad (4.55)$$

The trace part of (4.55) leads to

$$\alpha_j(\xi_j) = \frac{1}{4n}g((\mathcal{L}_{\xi_j}I_i)I_ke_a, e_a) \qquad (4.56)$$

while the trace-free part of (4.55) together with (4.56) determine completely the tensor U by

$$U(X,Y) = \frac{1}{4}g((\mathcal{L}_{\xi_j}I_i)X, I_kY) - \frac{1}{4}g((\mathcal{L}_{\xi_j}I_i)I_kX, Y)$$
$$+ \frac{1}{2n}g((\mathcal{L}_{\xi_j}I_i)I_ke_a, e_a)g(X,Y). \qquad (4.57)$$

Note that (4.57) shows that U belongs to the [3]-component because of the identity (4.26).

We determine the function $\lambda = T(\xi_i, \xi_j, \xi_k)$ inserting (4.56) into (4.42)

$$\lambda = d\eta_i(\xi_j, \xi_k) - d\eta_j(\xi_k, \xi_i) - d\eta_k(\xi_i, \xi_j) - \frac{1}{2n}g((\mathcal{L}_{\xi_j}I_i)I_ke_a, e_a). \qquad (4.58)$$

Finally, we calculate

$$(\mathcal{L}_{\xi_s}\omega_i)(\xi_t, I_iX) = -\omega_i([\xi_s, \xi_t], I_iX) = T(\xi_s, \xi_t, X). \qquad (4.59)$$

This completes the proof of the Biquard theorem. $\qquad\square$

Applying (4.14), we obtain from the proof of the preceeding theorem the next Corollary.

Corollary 4.2.8. *The Biquard and the Levi-Civita connections are connected by*

$$g(\nabla_{\xi_i}X, Y) = g(\nabla^g_{\xi_i}X, Y) + U(I_iX, Y) - \omega_i(X, Y),$$

$$g(\nabla_X\xi_i, Y) = g(\nabla^g_X\xi_i, Y) + \frac{1}{4}(T^0(I_i, X, Y) + T^0(X, I_iY)) - \omega_i(X, Y),$$

$$g(\nabla_A\xi_i, \xi_j) = g(\nabla^g_A\xi_i, \xi_j) - \frac{1}{2}T(\xi_i, \xi_j, A).$$

4.3 The curvature of the Biquard connection

The main purpose of this section is to show that the curvature of the Biquard connection is completely determined by its restriction to H and the torsion.

Let $R = [\nabla, \nabla] - \nabla_{[\ ,\ ]}$ be the curvature tensor of ∇. The curvature operator R_{BC} preserves the qc structure on M since ∇ preserves it. In particular R_{BC} preserves the distributions H and V, the quaternionic structure $\mathbb{Q}$ on H and the $(2,1)$ tensor φ. Moreover, the action of R_{BC} on V is completely determined by its action on H,

$$R_{BC}\xi_s = \varphi^{-1}([R_{BC}, I_s]).$$

Thus, we may regard R_{BC} on H as an endomorphism of H and we have $R_{BC} \in sp(n) \oplus sp(1)$.

As usual, we write $R(A, B, C, D) = g(R_{A,B}C, D)$.

Definition 4.3.1. The Ricci 2-forms ρ_i are defined by

$$\rho_i(B, C) = \frac{1}{4n} R(B, C, e_a, I_i e_a).$$

We decompose the curvature on H into $sp(n) \oplus sp(1)$-parts. Let $R^0_{BC} \in sp(n)$ denote the $sp(n)$-component.

Lemma 4.3.2.

a) *The curvature of the Biquard connection decomposes on H as follows*

$$R_{BC}X = R^0_{BC}X + \rho_1(B, C)I_1X + \rho_2(B, C)I_2X + \rho_3(B, C)I_3X,$$

$$R_{BC}I_iX - I_iR_{BC}X = 2(-\rho_j(B, C)I_kX + \rho_k(B, C)I_jX), \qquad (4.60)$$

$$\rho_i(B, C) = \frac{1}{2}(d\alpha_i + \alpha_j \wedge \alpha_k)(B, C), \qquad (4.61)$$

where the connection 1-forms α_s are determined in (4.16), (4.36) *and* (4.42).

b) *The curvature of the Biquard connection on V is determined by*

$$R(B, C, \xi_i, \xi_j) = 2\rho_k(B, C). \qquad (4.62)$$

Proof. The first two identities follow directly from the definitions. Using (4.10), we calculate that on H we have

$$R_{BC}I_i - I_iR_{BC} = \nabla_B\nabla_C I_i - \nabla_c\nabla_B I_i - \nabla_{[B,C]}I_i$$

$$= \nabla_B(\alpha_k(C)I_j - \alpha_j(C)I_k) - \nabla_C(\alpha_k(B)I_j - \alpha_j(B)I_k)$$

$$- (\alpha_k([B,C])I_j - \alpha_j([B,C])I_k)$$

$$= -(d\alpha_j + \alpha_k \wedge \alpha_i)(B,C)I_k + (d\alpha_k + \alpha_i \wedge \alpha_j)(B,C)I_j.$$

Now (4.61) follows from (4.60).

Similarly, using (4.10) and (4.61), we obtain

$$R(B, C)\xi_i = -(d\alpha_j + \alpha_k \wedge \alpha_i)(B,C)\xi_k + (d\alpha_k + \alpha_i \wedge \alpha_j)(B,C)\xi_j$$

$$= -2\rho_j(B,C)\xi_k + \rho_k(B,C)\xi_j.$$

$\square$

Definition 4.3.3. The quaternionic contact Ricci tensor (*qc-Ricci tensor* for short) and the qc-scalar curvature *Scal* of the Biquard connection are defined by

$$Ric(B, C) = R(e_a, B, C, e_a), \qquad Scal = Ric(e_a, e_a). \qquad (4.63)$$

It is known, cf. [24], that the qc-Ricci tensor restricted to H is symmetric. In addition, there are six Ricci-type tensors ζ_s, τ_s defined in [94] as follows

$$\begin{aligned}
\zeta_s(B,C) &= \frac{1}{4n} R(e_a, B, C, I_s e_a), \\
\tau_s(B,C) &= \frac{1}{4n} R(e_a, I_s e_a, B, C).
\end{aligned} \tag{4.64}$$

In fact all Ricci-type contractions evaluated on the horizontal space H are determined by the components of the torsion endomorphism and the qc-scalar curvature [94]. We begin with analysis of the Bianchi identity.

4.3.1   The first Bianchi identity and Ricci tensors

In this section we describe the horisontal Ricci tensors in terms of the torsion endomorphism of the Biquard connection and qc-scalar curvature thanks to Lemma 4.2.6, Lemma 4.2.7 and the first Bianchi identity.

Let $b(A, B, C)$ denote the Bianchi projector,

$$b(A,B,C) := \sum_{(A,B,C)} \left\{ (\nabla_A T)(B,C) + T(T(A,B),C) \right\}, \tag{4.65}$$

where $\sum_{(A,B,C)}$ denotes the cyclic sum over the three tangent vectors. With this notation the first Bianchi identity reads as follows

$$\sum_{(A,B,C)} \left\{ R(A,B,C,D) \right\} = g\Big(b(A,B,C),D\Big) = b(A,B,C,D). \tag{4.66}$$

In what follows we need the well known formula interchanging the places of the first couple and the second couple arguments of the curvature tensor, valid for any metric connection having non-trivial torsion.

The curvature of a linear connection preserving the metric is skew-symmetric with respect to the last two arguments, $R(A,B,C,D) = -R(A,B,D,C)$. One checks directly that the first Bianchi identity (4.66) yields

Lemma 4.3.4. *The curvature R of a metric connection with torsion T satisfies*

$$\begin{aligned}
\Big(R(A,B,C,D) - R(C,D,A,B) \Big) &= \frac{1}{2} b(A,B,C,D) + \frac{1}{2} b(B,C,D,A) \\
&\quad - \frac{1}{2} b(A,C,D,B) - \frac{1}{2} b(A,B,D,C). \tag{4.67}
\end{aligned}$$

The next result was originally proved in [94] and stated in the form below in [97].

Theorem 4.3.5. *On a* $(4n + 3)$-*dimensional qc manifold, the horizontal Ricci tensors Ric and* $\zeta_s(X, I_sY)$ *are symmetric and the horizontal Ricci tensors* $\rho_s(X, I_sY), \tau_s(X, I_sY)$ *are symmetric (1,1) tensors with respect to* I_s *and the next formulas hold*

$$Ric(X, Y) = \frac{Scal}{4n}g(X, Y)$$

$$+ (2n + 2)T^0(X, Y) + (4n + 10)U(X, Y); \quad (4.68)$$

$$\rho_s(X, I_sY) = -\frac{Scal}{8n(n + 2)}g(X, Y)$$

$$- \frac{1}{2}\Big[T^0(X, Y) + T^0(I_sX, I_sY)\Big] - 2U(X, Y); \quad (4.69)$$

$$\tau_s(X, I_sY) = -\frac{Scal}{8n(n + 2)}g(X, Y)$$

$$- \frac{n + 2}{2n}\Big[T^0(X, Y) + T^0(I_sX, I_sY)\Big]; \quad (4.70)$$

$$\zeta_s(X, I_sY) = \frac{Scal}{16n(n + 2)}g(X, Y)$$

$$+ \frac{2n + 1}{4n}T^0(X, Y) + \frac{1}{4n}T^0(I_sX, I_sY) + \frac{2n + 1}{2n}U(X, Y); \quad (4.71)$$

$$Scal = -8n(n + 2)g(T(\xi_1, \xi_2), \xi_3); \quad (4.72)$$

$$T(\xi_i, \xi_j) = -\frac{Scal}{8n(n + 2)}\xi_k - [\xi_i, \xi_j]_H; \quad (4.73)$$

$$T(\xi_i, \xi_j, I_kX) = \rho_k(I_jX, \xi_i) = -\rho_k(I_iX, \xi_j) = \omega_k([\xi_i, \xi_j], X); \quad (4.74)$$

$$\rho_i(X, \xi_i) = -\frac{X(Scal)}{32n(n + 2))}$$

$$+ \frac{1}{2}\left(-\rho_i(\xi_j, I_kX) + \rho_j(\xi_k, I_iX) + \rho_k(\xi_i, I_jX)\right); \quad (4.75)$$

$$\rho_i(\xi_i, \xi_j) + \rho_k(\xi_k, \xi_j) = \frac{1}{16n(n + 2)}\xi_j(Scal). \quad (4.76)$$

For $n = 1$ *the above formulas hold with* $U = 0$.

Proof. Since ∇ preserves the splitting $H \oplus V$, the first Bianchi identity (4.66) and (4.62) together with (4.11) and (4.34) imply

$$2\rho_i(X,Y) = R(X,Y,\xi_j,\xi_k) = b(X,Y,\xi_j,\xi_k) = (\nabla_{\xi_j}T)(X,Y,\xi_k)$$
$$+ T(T(X,Y),\xi_j),\xi_k) + T(T(Y,\xi_j),X),\xi_k) + T(T(\xi_j,X),Y),\xi_k)$$
$$= 2(\nabla_{\xi_j}\omega_k)(X,Y) - 2T(X,Y,\nabla_{\xi_j}\xi_k)$$
$$+ 2\omega_i(X,Y)T(\xi_i,\xi_j,\xi_k) + 2\omega_k(T(\xi_j,X),Y) - 2\omega_k(T(\xi_j,Y),X), \quad (4.77)$$

where we used the fact that $T(\xi_s,X)$ is a horizontal vector field, $T(\xi,X) \in H$ to conclude the vanishing of terms of the type $(\nabla_A T)(X,\xi_j,\xi_k)$.

Applying (4.10) and (4.11) we deduce that the third line in (4.77) vanishes because

$$T(X,Y,\nabla_{\xi_j}\xi_k) = -\alpha_i(\xi_j)\omega_j(X,Y) + \alpha_j(\xi_j)\omega_i(X,Y) = (\nabla_{\xi_j}\omega_k)(X,Y).$$

With the help of (4.41), (4.20), Lemma 4.2.6 and Lemma 4.2.7 we obtain from (4.77) that

$$\rho_i(X,Y) = -\lambda\omega_i(X,Y) - T(\xi_j,X,I_kY) + T(\xi_j,Y,I_kX)$$
$$= -\lambda\omega_i(X,Y) - 2U(I_iX,Y) - T^0(\xi_j,X,I_kY) + T^0(\xi_j,Y,I_kX)$$
$$= -\lambda\omega_i(X,Y) - 2U(I_iX,Y) + \frac{1}{2}(T^0(X,I_iY) - T^0(I_iX,Y)). \quad (4.78)$$

In view of (4.60), we have

$$Ric(B,I_sY) + 4n\zeta_s(B,Y)$$
$$= R(e_a,B,I_sY,e_a) + R(e_a,B,Y,I_se_a)$$
$$= -2\rho_j(e_a,B)\omega_k(Y,e_a) + 2\rho_k(e_a,B)\omega_j(Y,e_a))$$
$$= 2\rho_j(B,I_kY) - 2\rho_k(B,I_jY). \quad (4.79)$$

Consequently, applying (4.78), we derive from (4.79) and the properties of the torsion listed in Lemmas 4.2.6 and 4.2.7 that

$$Ric(X,I_sY) + 4n\zeta_s(X,Y) = -4\lambda\omega_s(X,Y)$$
$$+ 4U(X,I_sY) + \frac{1}{2}\Big(T^0(X,I_sY) + T^0(I_sX,Y)\Big). \quad (4.80)$$

The first Bianchi identity (4.66) implies

$$4n(\tau_s(X,Y) + 2\zeta_s(X,Y))$$
$$= R(e_a,I_se_a,X,Y) + R(X,e_a,I_se_a,Y) + R(I_se_a,X,e_a,Y)$$
$$= b(e_a,I_se_a,X,Y). \quad (4.81)$$

Finally, (4.67) yields

$$8n(\tau_s(X,Y) - \rho_s(X,Y)) = 2R(e_a, I_s e_a, X, Y) - 2R(X, Y, e_a, I_s e_a)$$
$$= b(e_a, I_s e_a, X, Y) - 2b(e_a, X, Y, I_s e_a) - b(e_a, I_s e_a, Y, X). \quad (4.82)$$

We consider the Bianchi projector b restricted to H. Since ∇ preserves the splitting $H \oplus V$, we have

$$X.g(T(Y,Z), W) = 0, \quad g(T(\nabla_X Y, Z), W) = 0.$$

Consequently, $(\nabla_X T)(Y, Z, W) = 0$. Applying (4.11), Lemmas 4.2.6 and 4.2.7 we calculate the bianchi projector $b_{|H}$ to be

$$b(X, Y, Z, W) = 2 \sum_{(X,Y,Z)} \sum_{s} \omega_s(X, Y) T(\xi_s, Z, W)$$

$$= \sum_{(X,Y,Z)} \sum_{s} \omega_s(X, Y) \left(2U(I_s Z, W) - \frac{1}{2} T^0(I_s Z, W) - \frac{1}{2} T^0(Z, I_s W) \right).$$
$$(4.83)$$

Applying Lemmas 4.2.6 and 4.2.7 to (4.83), we obtain after some standard calculations that

$$b(e_a, X, Y, I_s e_a) = b(X, Y, e_a, I_s e_a)$$
$$= 4U(I_s X, Y) + 2T^0(I_s X, Y) - 2T^0(X, I_s Y). \quad (4.84)$$

$$b(e_a, I_s e_a, X, Y) = (8n + 4)U(I_s X, Y)$$
$$- (2n + 2)T^0(I_s X, Y) - (2n - 2)T^0(X, I_s Y). \quad (4.85)$$

Substitute (4.84) and (4.85) into (4.82) and use (4.78) to conclude

$$\tau_s(X, Y) = -\lambda\omega_s(X, Y) + \frac{n+2}{2n}\left[T^0(X, I_s Y) - T^0(I_s X, Y) \right]. \quad (4.86)$$

Similarly, inserting (4.86) and (4.85) into (4.81) we find

$$\zeta_s(X, Y) = \frac{1}{2}\lambda\omega_s g(X, Y)$$
$$- \frac{2n+1}{4n}T^0(X, I_s Y) + \frac{1}{4n}T^0(I_s X, Y) - \frac{2n+1}{2n}U(X, I_s Y). \quad (4.87)$$

The equality (4.87) together with (4.80) yield

$$Ric(X, I_s Y) = -2(n+2)\lambda\omega_s(X, Y)$$
$$+ (2n+2)T^0(X, I_s Y) + (4n + 10)U(X, I_s Y). \quad (4.88)$$

Take the trace in (4.88) and use Lemmas 4.2.6 and 4.2.7 to conclude

$$\lambda = \frac{Scal}{8n(n+2)} \tag{4.89}$$

Insert (4.89) into (4.88), (4.87) and (4.86) to obtain (4.68), (4.71) and (4.70). The equalities (4.89) together with (4.41) and (4.34) imply (4.72) and (4.73).

Since ∇ preserves the splitting $H \oplus V$, the first Bianchi identity, (4.66) and (4.60) imply

$$2\rho_k(\xi_j, X) = R(\xi_j, X, \xi_i, \xi_j)$$
$$= \sum_{\xi_i, \xi_j, X} \{(\nabla_{\xi_i} T)(\xi_j, X, \xi_j) + T(T(\xi_i, \xi_j), X, \xi_j)\}$$
$$= (\nabla_X T)(\xi_i, \xi_j, \xi_j) + T(T(\xi_i, \xi_j), X, \xi_j)$$
$$= T(-[\xi_i, \xi_j]_H, X, \xi_j) = g([[\xi_i, \xi_j]_H, X], \xi_j) = -d\eta_j([\xi_i, \xi_j]_H, X)$$
$$= -2\omega_j([\xi_i, \xi_j], X) = -2T(\xi_i, \xi_j, I_j X), \quad (4.90)$$

where we used (4.73) and (4.10) for the third equality and (4.73) to establish the last line. This proves (4.74).

Similarly, using the just proved (4.74), we obtain

$$2\rho_i(X, \xi_i) + 2\rho_j(X, \xi_j) = R(X, \xi_i, \xi_j, \xi_k) + R(\xi_j, X, \xi_i, \xi_k)$$
$$= \sum_{\xi_i, \xi_j, X} \{(\nabla_{\xi_i} T)(\xi_j, X, \xi_k) + T(T(\xi_i, \xi_j), X, \xi_k)\}$$
$$= (\nabla_X T)(\xi_i, \xi_j, \xi_k) + T(T(\xi_i, \xi_j), X, \xi_k) = -\frac{X(Scal)}{8n(n+2)} - 2\omega_k([\xi_i, \xi_j], X)$$
$$= -\frac{X(Scal)}{8n(n+2)} - 2\rho_k(I_j X, \xi_i), \quad (4.91)$$

where we used (4.73) and (4.34) in the third line.
Take the cyclic permutations of the indices i, j, k in (4.91), summing up the first two obtained equalities and subtracting the third one, we obtain (4.75).

Since ∇ preserves the splitting $H \oplus V$, the first Bianchi identity , (4.74) (4.73) and (4.34) imply

$$-2(\rho_i(\xi_i, \xi_j) + \rho_k(\xi_k, \xi_j)) = \sum_{\xi_i, \xi_j, \xi_k} R(\xi_i, \xi_j, \xi_k, \xi_j)$$
$$= \sum_{\xi_i, \xi_j, \xi_k} \{(\nabla_{\xi_i} T)(\xi_j, \xi_k, \xi_j) + T(T(\xi_i, \xi_j), \xi_k, \xi_j)\}$$
$$= -\frac{1}{8n(n+2)} \xi_j(Scal)$$

which is precisely (4.76). This completes the proof of the theorem. $\qquad\square$

We have two straightforward consequences of the preceding proof.

Corollary 4.3.6. *The $sp(1)$-connection 1-forms α_s of the Biquard connection are given by*

$$\alpha_i(X) = d\eta_k(\xi_j, X) = -d\eta_j(\xi_k, X); \qquad (4.92)$$

$$\alpha_i(\xi_s) = d\eta_s(\xi_j, \xi_k) - \delta_{is}\frac{Scal}{16n(n+2)}$$

$$- \frac{1}{2}\delta_{is}\Big(d\eta_1(\xi_2, \xi_3) + d\eta_2(\xi_3, \xi_1) + d\eta_3(\xi_1, \xi_2)\Big). \quad (4.93)$$

Proof. The equality (4.92) is a combination of (4.16) and (4.34) and it was proved in [24]. Substitute (4.89) into (4.42) to get (4.93). $\qquad\square$

Taking the suitable traces and comparing the $Sp(n)Sp(1)$-invariant parts in (4.68), (4.69), (4.70) and (4.71), we obtain at once the next Corollary.

Corollary 4.3.7.

a) *The qc-scalar curvature satisfies the equalities*

$$\frac{Scal}{2(n+2)} = \rho_i(I_ie_a, e_a) = \tau_i(I_ie_a, e_a) = -2\zeta_i(I_ie_a, e_a).$$

b) *The tensors T^0 determines the traceless [-1]-component of the horizontal Ricci-type tensors while the tensor U determines the traceless part of the [3]-component of the horizontal Ricci-type tensors. For example, (4.68) yields*

$$T^0 = \frac{1}{2n+2}Ric_{[-1]}, \qquad U = \frac{1}{4n+10}Ric_{[3][0]}.$$

4.3.2 Local structure equations of qc manifolds

The fundamental 2-forms ω_s of a qc structure are locally defined horizontal 2-forms. We define a global horizontal four form Ω whose exterior derivative contains the essential information about the torsion endomorphism of the Biquard connection provided the dimension of the manifold is grater than seven. The $Sp(n)Sp(1)$-invariant fundamental four form of a given qc manifold is defined globally on the horizontal distribution H by [98]

$$\Omega = \omega_1 \wedge \omega_1 + \omega_2 \wedge \omega_2 + \omega_3 \wedge \omega_3. \qquad (4.94)$$

First we derive the local structure equations of a qc structure in terms of the $sp(1)$-connection forms of the Biquard connection and the qc scalar curvature.

Proposition 4.3.8. *Let $(M^{4n+3}, \eta, \mathbb{Q})$ be a $(4n+3)$- dimensional qc manifold with qc scalar curvature Scal. Let $s = \frac{Scal}{8n(n+2)}$ be the normalized qc scalar curvature. The following equations hold*

$$2\omega_i = d\eta_i + \eta_j \wedge \alpha_k - \eta_k \wedge \alpha_j + s\eta_j \wedge \eta_k, \qquad (4.95)$$

$$d\omega_i = \omega_j \wedge (\alpha_k + s\eta_k) - \omega_k \wedge (\alpha_j + s\eta_j)$$
$$- \rho_k \wedge \eta_j + \rho_j \wedge \eta_k + \frac{1}{2}ds \wedge \eta_j \wedge \eta_k, \qquad (4.96)$$

$$d\Omega = \sum_{(ijk)} \left[2\eta_i \wedge (\rho_k \wedge \omega_j - \rho_j \wedge \omega_k) + ds \wedge \omega_i \wedge \eta_j \wedge \eta_k \right], \qquad (4.97)$$

where α_s are the $sp(1)$-connection 1-forms of the Biquard connection, ρ_s are the Ricci 2-forms and $\sum_{(ijk)}$ is the cyclic sum of even permutations of $\{1, 2, 3\}$.

Proof. A straightforward calculation using (4.92) and (4.93) gives the equivalence of (4.29) and (4.95). Taking the exterior derivative of (4.95), followed by an application of (4.95), (4.61) and (4.62) implies (4.96). The last formula, (4.97), follows from (4.96) and definition (4.94). □

The next result originally proved in [98] expresses the tensors T^0 and U in terms of the exterior derivative of the fundamental four form. We have

Theorem 4.3.9. *On a qc manifold of dimension $(4n+3) > 7$ we have the identities*

$$U(X, Y) =$$
$$- \frac{1}{16n} \left[d\Omega(\xi_i, X, I_k Y, e_a, I_j e_a) + d\Omega(\xi_i, I_i X, I_j Y, e_a, I_j e_a) \right] \qquad (4.98)$$

$$T^0(X, Y) =$$
$$\frac{1}{8(1-n)} \sum_{(ijk)} \left[d\Omega(\xi_i, X, I_k Y, e_a, I_j e_a) - d\Omega(\xi_i, I_i X, I_j Y, e_a, I_j e_a) \right]. \qquad (4.99)$$

Proof. Equation (4.97) together with the second equality in Theorem 4.3.5 yield

$$d\Omega(\xi_i, X, I_kY, e_a, I_je_a)$$
$$= 4(n-1)\rho_k^0(X, I_kY) + 2\rho_j^0(X, I_jY) - 2\rho_j^0(I_iX, I_kY), \quad (4.100)$$

where ρ_s^0 is the horizontal trace-free part of ρ_s given by

$$\rho_s^0(X, I_sY) = -\frac{1}{2}\Big[T^0(X, Y) + T^0(I_sX, I_sY)\Big] - 2U(X, Y). \quad (4.101)$$

A substitution of (4.101) in (4.100), combined with the properties of the tensors T^0 and U described in Lemmas 4.2.6 and 4.2.7 give

$$d\Omega(\xi_i, X, I_kY, e_a, I_je_a)$$
$$= -2(n-1)\Big[T^0(X, Y) + T^0(I_kX, I_kY)\Big] - 8nU(X, Y). \quad (4.102)$$

Applying again Lemmas 4.2.6 and 4.2.7 to (4.102), we see that U and T^0 satisfy (4.98) and (4.99), respectively which completes the proof. $\square$

The well-known Cartan formula applied for the fundamental four form gives

$$\mathbb{L}_{\xi_s}\Omega = \xi_s \lrcorner d\Omega + d(\xi_s \lrcorner \Omega) = \xi_s \lrcorner d\Omega,$$

since Ω is horizontal. The latter formula and Theorem 4.3.9 yield

Corollary 4.3.10. *If one of the Reeb vector fields preserves the fundamental four form on a qc manifold of dimension $(4n+3) > 7$ then $U = 0$ and the torsion endomorphism of Biquard connection is symmetric, $T_{\xi_s} = T_{\xi_s}^0$.*
If on a qc manifold of dimension $(4n+3) > 7$ each Reeb vector field preserves the fundamental four form, $\mathbb{L}_{\xi_s}\Omega = 0$ then the torsion endomorphism of Biquard connection vanishes, $T_{\xi_s} = T^0 = U = 0$.

4.3.3 The curvature tensor

The next result, originaly proved in [94] and [97] describes the curvature of the Biquard connection. We have

Theorem 4.3.11. *On a QC manifold the curvature of the Biquard connection satisfies the equalities:*

$$R(X,Y,Z,V) - R(Z,V,X,Y)$$

$$= 2\sum_{s=1}^{3}\Big[\omega_s(X,Y)U(I_sZ,V) - \omega_s(Z,V)U(I_sX,Y)\Big]$$

$$- 2\sum_{s=1}^{3}\Big[\omega_s(X,Z)T^0(\xi_s,Y,V) + \omega_s(Y,V)T^0(\xi_s,Z,X)\Big]$$

$$+ 2\sum_{s=1}^{3}\Big[\omega_s(Y,Z)T^0(\xi_sX,V) + \omega_s(X,V)T^0(\xi_s,Z,Y)\Big]. \quad (4.103)$$

The [3]-componenet of the horizontal curvature with respect to the first two arguments is given by

$$3R(X,Y,Z,V) - \sum_{s=1}^{3}R(I_sX,I_sY,Z,V))$$

$$= 2\Big[g(Y,Z)T^0(X,V) + g(X,V)T^0(Z,Y)\Big]$$

$$- 2\Big[g(Z,X)T^0(Y,V) + g(V,Y)T^0(Z,X)\Big]$$

$$- 2\sum_{s=1}^{3}\Big[\omega_s(Y,Z)T^0(X,I_sV) + \omega_s(X,V)T^0(Z,I_sY)\Big]$$

$$+ 2\sum_{s=1}^{3}\Big[\omega_s(Z,X)T^0(Y,I_sV) + \omega_s(V,Y)T^0(Z,I_sX)\Big]$$

$$+ \sum_{s=1}^{3}\Big[2\omega_s(X,Y)\big(T^0(Z,I_sV) - T^0(I_sZ,V)\big) - 8\omega_s(Z,V)U(I_sX,Y)\Big]$$

$$- \frac{Scal}{2n(n+2)}\sum_{s=1}^{3}\omega_s(X,Y)\omega_s(Z,V). \quad (4.104)$$

The curvature with respect to one vertical direction is given by

$$R(\xi_i, X, Y, Z) = -(\nabla_X U)(I_i Y, Z)$$
$$-\frac{1}{4}\Big[(\nabla_Y T^0)(I_i Z, X) + (\nabla_Y T^0)(Z, I_i X)\Big]$$
$$+\frac{1}{4}\Big[(\nabla_Z T^0)(I_i Y, X) + (\nabla_Z T^0)(Y, I_i X)\Big]$$
$$+\omega_j(X, Y)\rho_k(I_i Z, \xi_i) - \omega_k(X, Y)\rho_j(I_i Z, \xi_i)$$
$$-\omega_j(X, Z)\rho_k(I_i Y, \xi_i) + \omega_k(X, Z)\rho_j(I_i Y, \xi_i)$$
$$-\omega_j(Y, Z)\rho_k(I_i X, \xi_i) + \omega_k(Y, Z)\rho_j(I_i X, \xi_i). \quad (4.105)$$

The vertical part of the Biquard curvature satisfies the next equality

$$R(\xi_i, \xi_j, X, Y) = (\nabla_{\xi_i} U)(I_j X, Y) - (\nabla_{\xi_j} U)(I_i X, Y)$$
$$-\frac{1}{4}\Big[(\nabla_{\xi_i} T^0)(I_j X, Y) + (\nabla_{\xi_i} T^0)(X, I_j Y)\Big]$$
$$+\frac{1}{4}\Big[(\nabla_{\xi_j} T^0)(I_i X, Y) + (\nabla_{\xi_j} T^0)(X, I_i Y)\Big]$$
$$-(\nabla_X \rho_k)(I_i Y, \xi_i) - \frac{Scal}{8n(n+2)}T(\xi_k, X, Y)$$
$$-T(\xi_j, X, e_a)T(\xi_i, e_a, Y) + T(\xi_j, e_a, Y)T(\xi_i, X, e_a), \quad (4.106)$$

where the Ricci 2-forms are given by

$$3(2n+1)\rho_i(\xi_i, X) = \frac{1}{4}(\nabla_{e_a} T^0)(e_a, X) - \frac{3}{4}(\nabla_{e_a} T^0)(I_i e_a, I_i X)$$
$$-(\nabla_{e_a} U)(X, e_a) + \frac{2n+1}{16n(n+2)}X(Scal), \quad (4.107)$$

$$3(2n+1)\rho_i(I_k X, \xi_j) = -3(2n+1)\rho_i(I_j X, \xi_k)$$
$$= -\frac{(2n+1)(2n-1)}{16n(n+2)}X(Scal) + 2(n+1)(\nabla_{e_a} U)(X, e_a)$$
$$+\frac{4n+1}{4}(\nabla_{e_a} T^0)(e_a, X) + \frac{3}{4}(\nabla_{e_a} T^0)(I_i e_a, I_i X). \quad (4.108)$$

Proof. The first Bianchi identity (4.66), its consequence (4.67) together with the help of Lemma 4.2.6, Lemma 4.2.7 and (4.11) imply the identity (4.103) in a straightforward way.

Taking into account (4.103) and (4.60), the properties of the torsion listed in Lemmas 4.2.6 and 4.2.7, we find

$$R(X,Y,Z,V) - R(I_iX, I_iY, Z, V) =$$
$$2\omega_j(X,Y)\rho_j(Z,V) + 2\omega_k(X,Y)\rho_k(Z,V)$$
$$+ 4\omega_j(X,Y)U(I_jZ,V) + 4\omega_k(X,Y)U(I_kZ,V)$$
$$- 4\omega_j(Z,V)U(I_jX,Y) - 4\omega_k(Z,V)U(I_kX,Y) \quad (4.109)$$

$$+ 2\Big[g(Y,Z)T^0(\xi_i, I_iX, V) + g(X,V)T^0(\xi_i, I_iZ, Y)\Big]$$
$$- 2\Big[g(Z,X)T^0(\xi_i, I_iY, V) + g(V,Y)T^0(\xi_i, I_iZ, X)\Big]$$
$$+ 2\Big[\omega_i(Y,Z)T^0(\xi_i, X, V) + \omega_i(X,V)T^0(\xi_i, Z, Y)\Big]$$
$$- 2\Big[\omega_i(X,Z)T^0(\xi_i, Y, V) + \omega_i(Y,V)T^0(\xi_i, Z, X)\Big]$$
$$- \frac{1}{2}\Big[\omega_j(Y,Z)\Big(T^0(X, I_jV) - T^0(I_iX, I_kV)\Big)\Big]$$
$$- \frac{1}{2}\Big[\omega_k(Y,Z)\Big(T^0(X, I_kV) + T^0(I_iX, I_jV)\Big)\Big]$$
$$- \frac{1}{2}\Big[\omega_j(X,V)\Big(T^0(Y, I_jZ) - T^0(I_iY, I_kZ)\Big)\Big]$$
$$- \frac{1}{2}\Big[\omega_k(X,V)\Big(T^0(Y, I_kZ) + T^0(I_iY, I_jZ)\Big)\Big]$$
$$+ \frac{1}{2}\Big[\omega_j(X,Z)\Big(T^0(Y, I_jV) - T^0(I_iY, I_kV)\Big)\Big]$$
$$+ \frac{1}{2}\Big[\omega_k(X,Z)\Big(T^0(Y, I_kV) + T^0(I_iY, I_jV)\Big)\Big]$$
$$+ \frac{1}{2}\Big[\omega_j(Y,V)\Big(T^0(X, I_jZ) - T^0(I_iX, I_kZ)\Big)\Big]$$
$$+ \frac{1}{2}\Big[\omega_k(Y,V)\Big(T^0(X, I_kZ) + T^0(I_iX, I_jZ)\Big)\Big].$$

Now, equality (4.104) follows from (4.109) and Theorem 4.3.5.

Invoking (4.65) and applying (4.11) and Theorem 4.3.5 we have

$$b(\xi_i, X, Y, Z) = -(\nabla_X T)(\xi_i, Y, Z) + (\nabla_Y T)(\xi_i, X, Z)$$
$$+ 2\omega_j(X,Y)\rho_k(I_iZ, \xi_i) - 2\omega_k(X,Y)\rho_j(I_iZ, \xi_i)$$
$$= \frac{1}{4}(\nabla_X T^0)(I_iY, Z) + \frac{1}{4}(\nabla_X T^0)(Y, I_iZ) - (\nabla_X U)(I_iY, Z)$$
$$- \frac{1}{4}(\nabla_Y T^0)(I_iX, Z) - \frac{1}{4}(\nabla_Y T^0)(X, I_iZ) + (\nabla_Y U)(I_iX, Z)$$
$$+ 2\omega_j(X,Y)\rho_k(I_iZ, \xi_i) - 2\omega_k(X,Y)\rho_j(I_iZ, \xi_i), \quad (4.110)$$

where we used Lemma 4.2.6, Lemma 4.2.7 and the two equalities in (4.10) to pass from the second to the third equality.

A substitution of (4.110) in (4.67) implies (4.105).

If we take the trace in (4.105) and apply (4.73) and (4.72) we come to

$$n\rho_i(\xi_i, X) + \frac{1}{2}\rho_k(I_j X, \xi_i) + \frac{1}{2}\rho_j(I_i X, \xi_k)$$
$$= \frac{1}{8}(\nabla_{e_a}T^0)(e_a, X) - \frac{1}{8}(\nabla_{e_a}T^0)(I_i e_a, I_i X). \quad (4.111)$$

Summing (4.111) and (4.75), we obtain

$$(n+1)\rho_i(\xi_i, X) + \frac{1}{2}\rho_i(I_k X, \xi_j)$$
$$= \frac{1}{8}(\nabla_{e_a}T^0)(e_a, X) - \frac{1}{8}(\nabla_{e_a}T^0)(I_i e_a, I_i X) + \frac{X(Scal)}{32n(n+2)}. \quad (4.112)$$

The second Bianchi identity

$$\sum_{(A,B,C)} \left\{ (\nabla_A R)(B, C, D, E) + R(T(A, B), C, D, E) \right\} = 0 \quad (4.113)$$

and (4.11) give

$$\sum_{(X,Y,Z)} \left[(\nabla_X R)(Y, Z, V, W) + 2\sum_{s=1}^{3} \omega_s(X, Y)R(\xi_s, Z, V, W) \right] = 0. \quad (4.114)$$

Letting $X = e_a$, $Y = I_i e_a$ in (4.114) we find

$$(\nabla_{e_a}R)(I_i e_a, Z, V, W) + 2n(\nabla_Z \tau_i)(V, W) + 2(2n-1)R(\xi_i, Z, V, W)$$
$$+ 2R(\xi_j, I_k Z, V, W) - 2R(\xi_k, I_j Z, V, W) = 0. \quad (4.115)$$

After taking the trace in (4.115) and applying the formulas in Theorem 4.3.5 we come to

$$(2n-1)\rho_i(\xi_i, X) - 2\rho_i(I_k X, \xi_j) = \frac{2n-1}{16n(n+2)}X(Scal)$$
$$- \frac{1}{4}\left[(\nabla_{e_a}T^0)(e_a, X) + (\nabla_{e_a}T^0)(I_i e_a, I_i X) \right] - (\nabla_{e_a}U)(X, e_a). \quad (4.116)$$

Now, (4.112) and (4.116) yield (4.107).

Finally, from (4.65) and the fact that ∇ preserves the splitting $H \oplus V$, we get

$$R(\xi_i, \xi_j, X, Y) = (\nabla_{\xi_i}T)(\xi_j, X, Y) - (\nabla_{\xi_j}T)(\xi_i, X, Y) + (\nabla_X T)(\xi_i, \xi_j, Y)$$
$$- \frac{Scal}{8n(n+2)}T(\xi_k, X, Y) + T(T(\xi_j, X), \xi_i, Y)$$
$$- T(T(\xi_j, X), \xi_i, Y), \quad (4.117)$$

where we applied (4.73). The third term in the second line can be evaluated using (4.74), and (4.10) as follows

$$(\nabla_X T)(\xi_i, \xi_j, Y)$$
$$= -X(\rho_k(I_i Y, \xi_i)) + \alpha_j(X)\rho_i(I_j Y, \xi_j) - \alpha_i(X)\rho_j(I_i Y, \xi_i). \quad (4.118)$$

On the other hand, we calculate using (4.10) that

$$(\nabla_X \rho_k)(I_i Y, \xi_i) = X(\rho_k(I_i Y, \xi_i))$$
$$+ \alpha_i(X)\rho_j(I_i Y, \xi_i) - \alpha_j(X)\rho_i(I_i Y, \xi_i)$$
$$+ \alpha_j(X)\rho_k(I_k Y, \xi_i) + \alpha_j(X)\rho_k(I_i Y, \xi_k)$$
$$- \alpha_k(X)\rho_k(I_j Y, \xi_i) - \alpha_k(X)\rho_k(I_i Y, \xi_j)$$
$$= X(\rho_k(I_i Y, \xi_i)) + \alpha_i(X)\rho_j(I_i Y, \xi_i)$$
$$- \alpha_j(X)\Big[\rho_i(I_i Y, \xi_i) - \rho_k(I_i Y, \xi_k) - \rho_k(I_k Y, \xi_i)\Big]$$
$$= X(\rho_k(I_i Y, \xi_i)) + \alpha_i(X)\rho_j(I_i Y, \xi_i) - \alpha_j(X)\rho_i(I_j Y, \xi_j), \quad (4.119)$$

where we apply (4.74) to obtain the vanishing of the last line in the second equality and (4.75) to obtain the final equality. Compare (4.118) with (4.119) to conclude

$$(\nabla_X T)(\xi_i, \xi_j, Y) = -(\nabla_X \rho_k)(I_i Y, \xi_i). \quad (4.120)$$

Substitute (4.120) into (4.117) and evaluate the first two terms in (4.117) involving covariant derivative similarly as the first two terms in (4.110) one verifies that (4.106) holds.   □

As a consequence of Theorem 4.3.11, substituting (4.108) and (4.107) into (4.75), we obtain the next formula established in [[94], Theorem 4.8]

Theorem 4.3.12. The contracted Bianchi identity. *On a qc manifold of dimension $4n + 3$ the next formula holds*

$$(n - 1)(\nabla_{e_a} T^0)(e_a, X) + 2(n + 2)(\nabla_{e_a} U)(e_a, X)$$
$$- \frac{(n - 1)(2n + 1)}{8n(n + 2)} d(Scal)(X) = 0. \quad (4.121)$$

An application of Corollary (4.3.7) to (4.121) yields

Corollary 4.3.13. *On a $4n+3$-dimensional qc manifold the contracted Bianchi identity (4.121) has the form*

$$\frac{n - 1}{2(n + 1)}(\nabla_{e_a} Ric_{[-1]})(e_a, X) + \frac{n + 2}{2n + 5} \nabla_{e_a}(Ric_{[3][0]})(e_a, X)$$
$$- \frac{(n - 1)(2n + 1)}{8n(n + 2)} d(Scal)(X) = 0. \quad (4.122)$$

Take the trace in (4.106) and apply Lemma 4.2.6, Lemma 4.2.7 and (4.108) to conclude

Corollary 4.3.14. *On a qc manifold of dimension $4n + 3$ we have the formulas*

$$
4n\rho_k(\xi_i, \xi_j) = (\nabla_{e_a}\rho_k)(I_j e_a, \xi_i) + \frac{1}{4}||T^0||^2 - 2||U||^2
$$
$$
= \frac{2n+2}{3(2n+1)}(\nabla_{e_a e_b}U)(e_a, e_b) + \frac{4n+1}{12(2n+1)}(\nabla_{e_a e_b}T^0)(e_a, e_b)
$$
$$
+ \frac{1}{4(2n+1)}(\nabla_{e_a e_b}T^0)(I_k e_a, I_k e_b) - \frac{2n-1}{48n(n+2)}e_a e_a(Scal)
$$
$$
+ \frac{1}{4}||T^0||^2 - 2||U||^2,
$$

$$
4n\rho_i(\xi_i, \xi_j) = (\nabla_{e_b}\rho_k)(e_b, \xi_i)
$$
$$
= \frac{2n-1}{48n(n+2)}\nabla_{e_b I_j e_b}Scal - \frac{2n+2}{3(2n+1)}(\nabla_{e_b e_a}U)(I_j e_b, e_a)
$$
$$
- \frac{4n+1}{12(2n+1)}(\nabla_{e_b e_a}T^0)(I_j e_b, e_a) + \frac{3}{12(2n+1)}(\nabla_{e_b e_a}T^0)(I_i e_b, I_k e_a),
$$

$$
4n\rho_k(\xi_j, \xi_k) = (\nabla_{e_b}\rho_i)(e_b, \xi_k)
$$
$$
= -\frac{2n-1}{48n(n+2)}\nabla_{e_b I_j e_b}Scal + \frac{2n+2}{3(2n+1)}(\nabla_{e_b e_a}U)(I_j e_b, e_a)
$$
$$
+ \frac{4n+1}{12(2n+1)}(\nabla_{e_b e_a}T^0)(I_j e_b, e_a) + \frac{3}{12(2n+1)}(\nabla_{e_b e_a}T^0)(I_k e_b, I_i e_a).
$$

4.3.4 The flat model - The qc Heisenberg group

The quaternionic Heisenberg group $G\,(\mathbb{H})$ with its standard left invariant quaternionic contact structure is the simplest example of a qc manifold. The Biquard connection coincides with the flat left-invariant connection on the group and $G\,(\mathbb{H})$ constitute the flat model of the quaternionic contact geometry.

We use the following model of the quaternionic Heisenberg group $G\,(\mathbb{H})$. Define $G\,(\mathbb{H}) = \mathbb{H}^n \times \operatorname{Im}\mathbb{H}$ with the group law given by

$$
(q', \omega') = (q_o, \omega_o) \circ (q, \omega) = (q_o + q, \omega + \omega_o + 2\operatorname{Im} q_o \bar{q}),
$$

where $q,\ q_o \in \mathbb{H}^n$ and $\omega, \omega_o \in \operatorname{Im}\mathbb{H}$.

In coordinates, with the obvious notation, the multiplication formula is

$$
\begin{aligned}
t'^a &= t^a + t_o^a, \quad x'^a = x^a + x_o^a, \\
y'^a &= y^a + x_o^a, \quad z'^a = z^a + z_o^a, \\
x' &= x + x_o + 2(x_o^a t^a - t_o^a x^a + z_o^a y^a - y_o^a z^a) \\
y' &= y + y_o + 2(y_o^a t^a - z_o^a x^a - t_o^a y^a + x_o^a z^a) \\
z' &= z + z_o + 2(z_o^a t^a + y_o^a x^a - x_o^a y^a - t_o^a z^a).
\end{aligned}
\tag{4.123}
$$

A basis of left invariant horizontal vector fields

$$
T_a, \quad X_a = I_1 T_a, \quad Y_a = I_2 T_a, \quad Z_a = I_3 T_a
$$

is given by

$$
\begin{aligned}
T_a &= \frac{\partial}{\partial t_\alpha} + 2x^a \frac{\partial}{\partial x} + 2y^a \frac{\partial}{\partial y} + 2z^a \frac{\partial}{\partial z} \\
X_a &= \frac{\partial}{\partial x_\alpha} - 2t^a \frac{\partial}{\partial x} - 2z^a \frac{\partial}{\partial y} + 2y^a \frac{\partial}{\partial z} \\
Y_a &= \frac{\partial}{\partial y_\alpha} + 2z^a \frac{\partial}{\partial x} - 2t^a \frac{\partial}{\partial y} - 2x^a \frac{\partial}{\partial z} \\
Z_a &= \frac{\partial}{\partial z_\alpha} - 2y^a \frac{\partial}{\partial x} + 2x^a \frac{\partial}{\partial y} - 2t^a \frac{\partial}{\partial z} .
\end{aligned}
\tag{4.124}
$$

The central (vertical) vector fields ξ_1, ξ_2, ξ_3 are described as follows

$$
\xi_1 = 2\frac{\partial}{\partial x} \quad \xi_2 = 2\frac{\partial}{\partial y} \quad \xi_3 = 2\frac{\partial}{\partial z} .
\tag{4.125}
$$

A small calculation shows the following commutator relations

$$
[I_j T_a, T_a] = 2\xi_j \qquad [I_j T_a, I_i T_a] = 2\xi_k.
\tag{4.126}
$$

The standard quaternionic contact form $\tilde{\Theta} = (\tilde{\Theta}_1, \tilde{\Theta}_2, \tilde{\Theta}_3)$ is

$$
2\tilde{\Theta} = d\omega - q' \cdot d\bar{q}' + dq' \cdot \bar{q}',
\tag{4.127}
$$

which, in coordinates reads

$$
\begin{aligned}
\tilde{\Theta}_1 &= \frac{1}{2} dx - x^a dt^a + t^a dx^a - z^a dy^a + y^a dz^a \\
\tilde{\Theta}_2 &= \frac{1}{2} dy - y^a dt^a + z^a dx^a + t^a dy^a - x^a dz^a \\
\tilde{\Theta}_3 &= \frac{1}{2} dz - z^a dt^a - y^a dx^a + x^a dy^a + t^a dz^a.
\end{aligned}
$$

The Biquard connection coincides with the flat left-invariant connection on $G\,(\mathbb{H})$ and the described horizontal and vertical vector fields are parallel with respect to the Biquard connection and constitute an orthonormal basis of the tangent space.

More precisely, an application of Theorem 4.3.11 gives a possibility to characterize locally the quaternionic Heisenberg group as the unique quaterninic contact manifold with flat horizontal curvature of the Biquard connection. The first part of the next theorem is established in [94] and the second was completed in [97].

Theorem 4.3.15. *A quaternionic contact manifold is locally isomorphic to the quaternionic Heisenberg group exactly when the curvature of the Biquard connection restricted to H vanishes, $R_{|H} = 0$.*

Proof. First we show that if the whole curvature vanishes than the quaternionic contact manifold is locally isomorphic to the quaternionic Heisenberg group.

Suppose that the Biquard connection ∇ is flat. Then all Ricci type tensors vanish and Theorem 4.3.5 shows that

$$T^0 = U = 0 \quad \Longleftrightarrow \quad T(\xi, X, Y) = 0, \quad \xi \in V.$$

We conclude from (4.74) that

$$T(\xi_i, \xi_j, I_k X) = \omega_k([\xi_i, \xi_j], X) = \rho_k(I_j X, \xi_i) = 0.$$

In particular the vertical distribution V is involutive.

In addition, applying (4.73) with $Scal = 0$, we finally obtain

$$T(\xi_i, \xi_j, \xi_k) = 0.$$

All that, combined with (4.34) show the torsion tensor of the Biquard connection is different from zero only on H, i.e. the whole torsion is given by (4.11).

Using (4.10) it is easy to verify from (4.11) the validity of the next

Lemma 4.3.16. *On a qc manifold, the torsion tensor of the Biquard connection ∇ restricted to H is ∇-parallel, $(\nabla_A T)(X, Y) = 0$.*

In our case, Lemma 4.3.16 shows that the whole torsion is ∇-parallel. This combined with the flatness of the curvature, $R = 0$ and the first Bianchi identity (4.66) gives that the torsion T satisfies the Jacobi identity

$$T(T(A, B), C) + T(T(B, C), A) + T(T(C, A)B) = 0.$$

Hence, the manifold has a local Lie group structure T by the Lie theorems. The structure equations of this Lie group determined by (4.11) are

$d\eta_s = 2\omega_s$ which precisely are the structure equations of the quaternionic Heisenberg group. Hence, applying again the Lie theorem we conclude that the manifold has a local Lie group structure which is locally isomorphic to $G(\mathbb{H})$. In other words, there is a local diffeomorphism $\Phi : M \to G(\mathbb{H})$ such that $\eta = \Phi^*\Theta$, where Θ is the standard contact form on $G(\mathbb{H})$, see (4.127).

Now, in order to complete the proof it is sufficient to show that if the horizontal curvature is zero then the full curvature tensor vanishes. The condition $R_{|_H} = 0$ and Theorem 4.3.5 yield $T^0 = U = Scal = 0$. The latter equalities substituted into (4.107) and (4.108) give $\rho_s(\xi_s, X) = \rho_s(\xi_t, X) = 0$. Now, (4.105) and (4.106) imply

$$R(\xi, X, Y, Z) = R(\xi_i, \xi_j, X, Y) = 0. \tag{4.128}$$

Furthermore, (4.62) and definition 4.3.1 yield

$$R(X, Y, \xi_i, \xi_j) = 2\rho_k(X, Y) = 0, \quad R(X, \xi, \xi_i, \xi_j) = 2\rho_k(X, \xi) = 0.$$

Finally, a combination of (4.62) together with (4.128) gives

$$4nR(\xi_s, \xi_t, \xi_i, \xi_j) = 8n\rho_k(\xi_s, \xi_t) = 2R(\xi_s, \xi_t, e_a, I_k e_a) = 0,$$

which ends the proof. $\qquad\qquad\qquad\qquad\qquad\qquad\qquad\qquad\qquad\square$

Examples of flat qc structures. As a Lie group the quaternionic Heisenberg group $G(\mathbb{H})$ can be characterized by the following structure equations. Denote by $e^a, 1 \leq a \leq (4n + 3)$ the basis of the left invariant 1-forms, and by e^{ab} the wedge product $e^a \wedge e^b$. The $(4n + 3)$-dimensional quaternionic Heisenberg Lie algebra is the 2-step nilpotent Lie algebra defined by:

$$de^a = 0, \qquad 1 \leq a \leq 4n,$$
$$d\eta_1 = de^{4n+1} = 2(e^{12} + e^{34} + \cdots + e^{(4n-3)(4n-2)} + e^{(4n-1)4n}) = 2\omega_1,$$
$$d\eta_2 = de^{4n+2} = 2(e^{13} + e^{42} + \cdots + e^{(4n-3)(4n-1)} + e^{4n(4n-2)}) = 2\omega_2,$$
$$d\eta_3 = de^{4n+3} = 2(e^{14} + e^{23} + \cdots + e^{(4n-3)4n} + e^{(4n-2)(4n-1)}) = 2\omega_3.$$

The Biquard connection coincides with the flat left-invariant connection on $G(\mathbb{H})$. This flat qc structure on the quaternionic Heisenberg group $G(\mathbb{H})$ is (locally) the unique qc structure with flat Biquard connection according to Theorem 4.3.15. By a rotation of the 1-forms defining the horizontal space of $G(\mathbb{H})$ we obtain an equivalent qc-structure (with the same Biquard connection). It is possible to introduce a different not two step nilpotent group structure on $\mathbb{H}^n \times \mathrm{Im}\,\mathbb{H}$ with respect to which the rotated forms are left invariant (but not parallel!) [10]. Following is an explicit description of this construction in dimension seven which we take from [10].

Consider the seven dimensional quaternionic Heisenberg group. Since e^4 is closed we can write $e^4 = dx_4$, where x_4 is a global function on the manifold $\mathbb{H} \times \mathrm{Im}\,\mathbb{H}$. Now we can use this function to define a non-left-invariant qc structure on this manifold as follows. For each $c \in \mathbb{R}$, let

$$\gamma^1 = e^1, \quad \gamma^4 = e^4, \quad \gamma^7 = e^7$$

$$\gamma^2 = \sin(-cx_4)\,e^2 + \cos(-cx_4)\,e^3, \quad \gamma^3 = -\cos(-cx_4)\,e^2 + \sin(-cx_4)\,e^3,$$

$$\gamma^5 = \sin(-cx_4)\,e^5 + \cos(-cx_4)\,e^6, \quad \gamma^6 = -\cos(-cx_4)\,e^5 + \sin(-cx_4)\,e^6.$$

A direct calculation shows that for $c \neq 0$ the forms $\{\gamma^l,\ 1 \leq l \leq 7\}$ define a unique Lie algebra $\mathfrak{l}_0$ with the following structure equations

$$d\gamma^1 = 0, \quad d\gamma^2 = -c\gamma^{34}, \quad d\gamma^3 = c\gamma^{24}, \quad d\gamma^4 = 0,$$
$$d\gamma^5 = 2\gamma^{12} + 2\gamma^{34} + c\gamma^{46}, \quad d\gamma^6 = 2\gamma^{13} + 2\gamma^{42} - c\gamma^{45}, \quad (4.129)$$
$$d\gamma^7 = 2\gamma^{14} + 2\gamma^{23}.$$

In particular, $\mathfrak{l}_0$ is an indecomposable solvable Lie algebra. Let $e_l, 1 \leq l \leq 7$ be the left invariant vector fields dual to the 1-forms $\gamma^l, 1 \leq l \leq 7$. The (global) flat qc structure on $\mathbb{H} \times \mathrm{Im}\,\mathbb{H}$ can also be described as follows $\eta_1 = \gamma^5$, $\eta_2 = \gamma^6$, $\eta_3 = \gamma^7$, $H = span\{\gamma^1, \ldots, \gamma^4\}$, $\omega_1 = \gamma^{12} + \gamma^{34}$, $\omega_2 = \gamma^{13} + \gamma^{42}$, $\omega_3 = \gamma^{14} + \gamma^{23}$. It is straightforward to check from (4.129) that the vector fields $\xi_1 = e_5$, $\xi_2 = e_6$, $\xi_3 = e_7$ satisfy the Duchemin compatibility conditions (4.7) and therefore the Biquard connection exists and ξ_s are the Reeb vector fields.

Let $(L_0, \eta, \mathbb{Q})$ be the simply connected connected Lie group with Lie algebra $\mathfrak{l}_0$ equipped with the left invariant qc structure $(\eta, \mathbb{Q})$ defined above. Then, as a consequence of the above construction, the torsion endomorphism and the curvature of the Biquard connection are identically zero but the basis $\gamma_1, \ldots, \gamma_7$ is not parallel. The $Sp(1)$-connection 1-forms in the basis $\gamma^1, \ldots, \gamma^7$ are given by $\alpha_1 = 0$, $\alpha_2 = 0$, $\alpha_3 = c\gamma^4$.

4.4 qc-Einstein quaternionic contact structures

The aim of this section is to show that the vanishing of the torsion endomorphism of the Biquard connection implies that the qc-scalar curvature is constant. The Bianchi identities will have an important role in the analysis.

Definition 4.4.1. A quaternionic contact structure is *qc-Einstein* if the qc-Ricci tensor is trace-free,

$$Ric(X,Y) = \frac{Scal}{4n}g(X,Y).$$

The next result, originally proved in [94], (see also [98]) is important in stating and analyzing the Yamabe problem in quaternionic contact setting.

Theorem 4.4.2. *Let* $(M, g, \mathbb{Q})$ *be a quaternionic contact manifold of dimension* $(4n + 3)$. *Then*

 a). $(M, g, \mathbb{Q})$ *is a qc-Einstein if and only if the tensors* $T^0 = U = 0$, *i.e. the torsion endomorphism vanishes identically,* $T_\xi = 0, \xi \in V$.

 b). *On a qc-Einstein manifold of dimension bigger than seven the qc scalar curvature is constant,* $d(Scal) = 0$ *and the vertical distribution spanned by the Reeb vector fields is integrble,* $[\xi_s, \xi_t] \in V$.

 c). *If* $n > 1$ *then* $(M, g, \mathbb{Q})$ *is qc-Einstein if and only if the fundamental four form is closed,* $d\Omega = 0$.

Proof. The part a) of the assertion follows from Corollary 4.3.7.

The first part of b) is a consequence of the just established part a) and (4.121) since $n > 1$. Substitute $T^0 = U = d(Scal) = 0$ into (4.108) to conclude $\rho_i(\xi_j, X) = 0$ and compare this with (4.74) to establish the integrability of the vertical distribution V.

To proof c), assume $T^0 = U = 0$ and $n > 1$. Then Theorem 4.3.5 implies

$$\rho_s(X,Y) = -s\omega_s(X,Y), \quad \rho_s(\xi_t, X) = 0,$$
$$\rho_i(\xi_i, \xi_j) + \rho_k(\xi_k, \xi_j) = 0, \tag{4.130}$$

since $Scal$ is constant and the horizontal distribution is integrable. Using the just obtained identities in (4.130), we obtain from (4.97) that $d\Omega = 0$.

The converse of c) follows directly from Theorem 4.3.9 which completes the proof of the theorem. □

4.4.1 Examples of qc-Einstein structures

The 3-Sasakian Case. We have adopted the definition that a $4n + 3$-dimensional (pseudo) Riemannian manifold (M, g_M) of signature either $(4n+3, 0)$ or $(4n, 3)$ has a *3-Sasakian structure* if the cone metric $t^2 g_M + \epsilon dt^2$ on $M \times \mathbb{R}$ is a hyperkähler metric of signature $(4n+4, 0)$ (*positive 3-Sasakian structure,* $\epsilon = 1$) or (4n,4) (*negative 3-Sasakian structure,* $\epsilon = -1$), respectively, see [113; 162; 100] for the negative case. In other words, the cone metric has holonomy contained in $Sp(n + 1)$ (see [29]) or in $Sp(n, 1)$ (see [3]), respectively. We recall that an almost hyperhermitian structure (g, J_1, J_2, J_3) is hyper-Kähler if the almost complex structures J_s are parallel with respect to the Levi-Civita connection of the metric g.

We remind that, usually, a $4n + 3$-dimensional Riemannian manifold (M, g) is called 3-Sasakian only in the positive case, while the term pseudo 3-Sasakain is used in the negative case. However, we find it convenient to use the more general definition.

4.4.2 Cones over a quaternionic contact structure

In this subsection we describe explicitly two almost hyperhermitian structures on the (pseudo) metric cone over a quaternionic contact manifold M^{4n+3} of dimension $(4n + 3)$. One of the strictures is the usual Riemannian metric on the cone, so we focus on the second structure, which is pseudo-Riemannian of signature $(4n, 4)$. Let either $\epsilon = 1$ or $\epsilon = -1$. Consider the cone $N = M^{4n+3} \times \mathbb{R}^+$ as the product of two smooth manifolds equipped, in addition, with the following (pseudo-)Riemannian metric G_N and 2-forms F_i:

$$
\begin{aligned}
G_N &= t^2 g + \epsilon t^2 (\eta_1 \odot \eta_1 + \eta_2 \odot \eta_2 + \eta_3 \odot \eta_3) + \epsilon dt \odot dt \\
F_i^\epsilon &= t^2 \omega_i + \epsilon t^2 \eta_j \wedge \eta_k - t \eta_i \wedge dt,
\end{aligned}
\tag{4.131}
$$

where $\odot$ is the symmetric product. In the above we have also denoted with the same later both a tensor and its lift to a tensor on the tangent bundle of N. The above tensors determine the following $(1, 1)$ tensors on the cone N

$$
\begin{aligned}
\phi_{i|H} &= I_i, \qquad \phi_i \xi_j = \xi_k, \qquad \phi_i \xi_k = -\xi_j, \\
\phi_i \xi_i &= -\epsilon t \frac{\partial}{\partial t}, \qquad \phi_i \left(t \frac{\partial}{\partial t} \right) = -\xi_i.
\end{aligned}
\tag{4.132}
$$

We claim that the above structures define a (pseudo) almost hyperhermitian structures on N. The claim follows from the following identities relating the 2-forms, the pseudo-metric tensor and the almost complex structures:

$$G_N(\phi_i X, Y) = F_i^\epsilon(X, Y) \iff g(\phi_i X, Y) = g(I_i X, Y),$$

$$G_N(\phi_i X, \xi_s) = F_i^\epsilon(X, \xi_s) \iff \eta_s(\phi_i X) = 0,$$

$$G_N(\phi_i X, \frac{\partial}{\partial t}) = F_i^\epsilon(X, \frac{\partial}{\partial t}) = 0 \text{ trivially,}$$

$$G_N(\phi_i \frac{\partial}{\partial t}, \xi_s) = F_i^\epsilon(\frac{\partial}{\partial t}, \xi_s) \iff \eta_s\left(\phi_i(t\frac{\partial}{\partial t})\right) = \epsilon\delta_{is},$$

$$G_N(\phi_i \frac{\partial}{\partial t}, \frac{\partial}{\partial t}) = F_i^\epsilon(\frac{\partial}{\partial t}, \frac{\partial}{\partial t}) = 0 \text{ trivially,}$$

$$G_N(\phi_i \frac{\partial}{\partial t}, X) = F_i^\epsilon(\frac{\partial}{\partial t}, X) \iff g(\phi_i \frac{\partial}{\partial t}, X) = 0,$$

$$G_N(\phi_i \xi_s, \xi_t) = F_i^\epsilon(\xi_s, \xi_t) \iff \eta_t(\phi_i \xi_j) = \delta_{kt}, \eta_t(\phi_i \xi_k) = -\delta_{jt}, \eta_t(\phi_i \xi_i) = 0,$$

$$G_N(\phi_i \xi_i, \frac{\partial}{\partial t}) = F_i^\epsilon(\xi_i, \frac{\partial}{\partial t}) \iff dt(\phi_i \xi_i) = -\epsilon t,$$

$$G_N(\phi_i \xi_s, X) = F_i^\epsilon(\xi_s, X) \iff g(\phi_i \xi_s, X) = 0.$$

Let us remind that the Riemannian cone is the case $\epsilon = 1$, and the above construction turns into the "standard" almost hyperhermitian structure on N.

Conversely, consider the cone $N = M^{4n+3} \times \mathbb{R}^+$ over a smooth $(4n+3)$-dimensional manifold M^{4n+3} equipped with a (pseudo) almost hyperhermitian structure (G_N, ϕ_l) where G_N is a (pseudo) Riemannian metric of signature $(4n+4, 0)$ (resp. $(4n, 4)$) for $\epsilon = 1$ (resp. $\epsilon = -1$) and ϕ_l, $l = 1, 2, 3$, are three anti-commuting almost complex structures. The 1-form dt on $\mathbb{R}^+$ and the three almost complex structures are related to three 1-forms η_l on M^{4n+3} defined by $\eta_l = \epsilon \phi_l(\frac{1}{t}dt)$, where we used the same notation for both a tensor and its lift to a tensor on the tangent bundle of N identifying M with the slice $t = 1$ of N. We may write the metric G_N and the three Kähler 2-forms on N as in (4.131) where $g = G_{N|H}$ and $H = \cap_{l=1}^3 Ker \, \eta_l$. A qc structure on M^{4n+3} is defined by the three 1-forms η_l.

We apply Hitchin's theorem stating that an almost hyperhermitian structure is hyperkähler if and only if the three Kähler 2-forms are closed [90], which is valid with the same proof in the case of non-positive definite metrics.

It is straightforward to check from the second equation in (4.131) that

$$dF_i^\epsilon = tdt \wedge (2\omega_i + 2\epsilon\eta_j \wedge \eta_k - d\eta_i) + t^2 d(\omega_i + \epsilon\eta_j \wedge \eta_k). \qquad (4.133)$$

The equation (4.133) yields that the three Kähler forms on N are closed, $dF_s^\epsilon = 0$, exactly when

$$d\eta_i = 2\omega_i + 2\epsilon\eta_j \wedge \eta_k. \qquad (4.134)$$

Therefore the cone metric G_N is hyperkähler, i.e. M^{4n+3} is 3-Sasakian, if and only if (4.134) is fulfilled.

We use equations (4.134) to find

$$\xi_i \lrcorner d\eta_{j|H} = 0, \quad d\eta_i(\xi_j, \xi_k) = 2\epsilon, \quad d\eta_i(\xi_i, \xi_k) = d\eta_i(\xi_i, \xi_j) = 0.$$

This quaternionic contact structure satisfies the conditions (4.7) and therefore it admits the Biquard connection ∇. The $sp(1)$ connection 1-forms of the Biquard connection we calculate from (4.92) and (4.93) to be

$$\alpha_s = -\Big(\frac{Scal}{16n(n+2)} + \epsilon\Big)\eta_s.$$

The latter equality together with (4.61) yield

$$\rho_s(X, Y) = \frac{1}{2}d\alpha_s(X, Y) = -\Big(\frac{Scal}{16n(n+2)} + \epsilon\Big)\omega_s(X, Y) \qquad (4.135)$$

Comparing (4.135) with (4.69) we obtain that the torsion endomorphism of the Biquard connection of the corresponding quaternionic contact structure vanishes identically and the qc scalar curvature is constant depending on the dimension, $T = U = 0$, $Scal = 16n(n+2)\epsilon$, see [94]. More precisely, we conclude in a straightforward way applying Theorem 4.3.11 and Corollary 4.2.8 the following result established originally in [94].

Theorem 4.4.3. *Any 3-Sasakian manifold is a qc-Einstein with qc-scalar curvature given by*

$$Scal = 16n(n+2)\epsilon.$$

The structure equations of a 3-Sasakian manifolds are the equations (4.134).

The Ricci-type tensors are given by

$$\begin{aligned}
\rho_{t|H} &= \tau_{t|H} = -2\zeta_{t|H} = -2\epsilon\omega_t \\
Ric(\xi_s, X) &= \rho_s(X, \xi_t) = \zeta_s(X, \xi_t) = \rho_s(\xi_t, \xi_r) = 0.
\end{aligned} \qquad (4.136)$$

The curvature R of the Biquard connection is expressed in terms of the curvature of the Levi-Civita connection R^g as follows

$$R(X, Y, Z, W) = R^g(X, Y, Z, W)$$

$$+ \sum_{s=1}^{3} \Big[\omega_s(Y, Z)\omega_s(X, W) - \omega_s(X, Z)\omega_s(Y, W) - 2\omega_s(X, Y)\omega_s(Z, W)\Big]$$

$$0 = R(\xi, Y, Z, W) = R^g(\xi, Y, Z, W);$$
$$0 = R(\xi, \bar{\xi}, Z, W) = R^g(\xi, \bar{\xi}, Z, W);$$

$$R(X, Y, \xi, \bar{\xi}) =$$
$$-4\Big(\eta_1 \wedge \eta_2(\xi, \bar{\xi})\omega_3 + \eta_2 \wedge \eta_3(\xi, \bar{\xi})\omega_1 + \eta_3 \wedge \eta_1(\xi, \bar{\xi})\omega_2\Big)(X, Y),$$

where $\xi, \bar{\xi} \in V$.

There are known many examples of positive 3-Sasakian manifold of dimension $(4n + 3)$, see [28] and references therein for a nice overview of positive 3-Sasakian spaces. Certain $SO(3)$-bundles over quaternionic Kähler manifolds with negative scalar curvature constructed in [113; 162; 100; 3] supply examples of negative 3-Sasakian manifolds. A natural definite metric on the negative 3-Sasakian manifolds is constructed in [162; 100] changing the sign of the metric on the vertical $SO(3)$-factor. With respect to this metric the negative 3-Sasakian manifolds become an A-manifold and its Riemannian Ricci tensor has precisely two constant eigenvalues, the negative one equal to $-4n - 14$ of multiplicity $4n$ and a positive one equal to $4n + 2$ of multiplicity 3 [100] and the Riemannian scalar curvature is the negative constant $-16n^2 - 44n + 6$ [162; 100]. We recall that a Riemannian manifold is called an A-manifold if its Riemannian Ricci tensor satisfies $(\nabla_A^g Ric^g)(A, A) = 0$ [84]. Explicit examples of negative 3-Sasakian manifolds are constructed in [11]. We take the next two examples from [11].

Zero torsion qc structure Example 1. Denote $\{\tilde{e}^l, \ 1 \leq l \leq 7\}$ the basis of the left invariant 1-forms and consider the simply connected Lie group with Lie algebra L_1 defined by the following equations:

$$de^1 = 0, \quad de^2 = -e^{12} - 2e^{34} - \frac{1}{2}e^{37} + \frac{1}{2}e^{46},$$

$$de^3 = -e^{13} + 2e^{24} + \frac{1}{2}e^{27} - \frac{1}{2}e^{45},$$

$$de^4 = -e^{14} - 2e^{23} - \frac{1}{2}e^{26} + \frac{1}{2}e^{35} \tag{4.137}$$

$$de^5 = 2e^{12} + 2e^{34} - \frac{1}{2}e^{67}, \quad de^6 = 2e^{13} + 2e^{42} + \frac{1}{2}e^{57},$$

$$de^7 = 2e^{14} + 2e^{23} - \frac{1}{2}e^{56},$$

and $e_l, 1 \leq l \leq 7$ be the left invariant vector field dual to the 1-forms $e^l, 1 \leq l \leq 7$, respectively. We define a global qc structure on L_1 by setting

$$\eta_1 = e^5, \quad \eta_2 = e^6, \quad \eta_3 = e^7, \quad H = span\{e^1, \ldots, e^4\},$$
$$\omega_1 = e^{12} + e^{34}, \quad \omega_2 = e^{13} + e^{42}, \quad \omega_3 = e^{14} + e^{23}. \tag{4.138}$$

It is straightforward to check from (4.137) that the vector fields $\xi_1 = e_5$, $\xi_2 = e_6$, $\xi_3 = e_7$ satisfy the Duchemin compatibility conditions (4.7) and therefore the Biquard connection exists and ξ_s are the Reeb vector fields. The structure equations (4.137) easily imply (see e.g. [11]) that the torsion endomorphism of the Biquard connection is zero, $T^0 = U = 0$ and the qc scalar curvature is a negative constant, $Scal = -4n(n+2) = -12$.

Zero torsion qc structure Example 2. Consider the simply connected Lie group L_2 with Lie algebra defined by the equations:

$$de^1 = 0, \quad de^2 = -e^{12} + e^{34}, \quad de^3 = -\frac{1}{2}e^{13}, \quad de^4 = -\frac{1}{2}e^{14},$$

$$de^5 = 2e^{12} + 2e^{34} + e^{37} - e^{46} + \frac{1}{4}e^{67},$$

$$de^6 = 2e^{13} - 2e^{24} - \frac{1}{2}e^{27} + e^{45} - \frac{1}{4}e^{57}, \tag{4.139}$$

$$de^7 = 2e^{14} + 2e^{23} + \frac{1}{2}e^{26} - e^{35} + \frac{1}{4}e^{56}.$$

A global qc structure on L_2 is defined by (4.138). It is easy to check from (4.139) that the triple $\{\xi_1 = e_5, \xi_2 = e_6, \xi_3 = e_7\}$ form the Reeb vector fields satisfying (4.7) and therefore the Biquard connection do exists. It is easy to calculate from the structure equations (4.139) (see e.g. [11]), that the torsion endomorphism of the Biquard connection is zero, $T^0 = U = 0$ and the qc scalar curvature is a negative constant, $S = -2n(n+2) = -6$.

It turns out that the 3-Sasakian spaces are locally the only qc-Einstein manifolds, the fact established in [94]. We state the result in its whole generality but shall give here a proof only for dimensions bigger than seven. We have

Theorem 4.4.4. *Let* $(M^{4n+3}, \eta, \mathbb{Q})$ *be a* $4n + 3$-*dimensional qc manifold with non-zero qc scalar curvature Scal and let* ϵ *be the sign of the scalar curvature. For* $n > 1$ *the following conditions are equivalent*

a) $(M^{4n+3}, g, \mathbb{Q})$ *is qc-Einstein manifold;*

b) M^{4n+3} *is locally qc homothetic to a 3-Sasakian manifold, i.e., locally, there exists a* $SO(3)$-*matrix* Ψ *with smooth entries depending on an auxiliary parameter, such that, the local qc structure* $(\epsilon\frac{16n(n+2)}{Scal}\Psi \cdot \eta, \mathbb{Q})$ *is 3-Sasakian.*

A seven dimensionl qc manifold is qc-Einstein with constant non-zero qc scalar curvature if and only if it is locally qc homothetic to a 3-sasakian manifold in the sense of b).

Proof. The idea is the same as in the proof of Theorem 3.1 in [94]. Let $T^0 = U = 0$ and $n > 1$. Theorem 4.4.2 shows that the qc scalar curvature is constant and the vertical distribution is integrable. The qc structure $\eta' = \frac{16n(n+2)}{\epsilon Scal}\eta$ has normalized qc scalar curvature $s' = 2\epsilon$ and $d\Omega' = 0$ provided $Scal \neq 0$. For simplicity, we shall denote η' with η and, in fact, drop the $'$ everywhere.

In the first step of the proof we show that the Riemannian cone $N = M \times \mathbb{R}^+$ with the metric $g_N = t^2(g + \epsilon \sum_{s=1}^3 \eta_s \otimes \eta_s) + \epsilon dt \otimes dt$ has holonomy contained either in $Sp(n+1)$ or in $Sp(n,1)$ depending on whether ϵ is positive or negative, respectively. To this end we consider the following four form on N

$$F = F_1 \wedge F_1 + F_2 \wedge F_2 + F_3 \wedge F_3, \tag{4.140}$$

where the two forms F_s are defined by

$$F_i = t^2(\omega_i + \epsilon\eta_j \wedge \eta_k) + tdt \wedge \eta_i. \tag{4.141}$$

Applying (4.95), (4.96) and (4.97), we calculate from (4.141) that

$$
\begin{aligned}
dF_i &= tdt \wedge \left(2\omega_i + 2\epsilon\eta_j \wedge \eta_k - d\eta_i\right) + t^2 d(\omega_i + \epsilon\eta_j \wedge \eta_k) \\
&= t\,dt \wedge \left(2\omega_i + 2\epsilon\eta_j \wedge \eta_k + \eta_j \wedge \alpha_k - \eta_k \wedge \alpha_k + s\eta_j \wedge \eta_k\right) \\
&\quad + \epsilon t^2\left(2\omega_j + \eta_i \wedge \alpha_k\right) \wedge \eta_k - \epsilon t^2\left(2\omega_k - \eta_i \wedge \alpha_j\right) \wedge \eta_j \\
&\quad + t^2\left(\omega_j \wedge (\alpha_k + s\eta_k) - \omega_k \wedge (\alpha_j + s\eta_j)\right) \\
&\quad - t^2\left(\rho_k \wedge \eta_j - \rho_j \wedge \eta_k - \frac{1}{2}ds \wedge \eta_j \wedge \eta_k\right). \tag{4.142}
\end{aligned}
$$

A short computation, using (4.95), (4.96), (4.97) and (4.142), gives

$$
\begin{aligned}
\frac{1}{2}dF &= \sum_{s=1}^3 dF_s \wedge F_s = t^3 dt \wedge \sum_{(ijk)}\left[2\epsilon\omega_i \wedge \eta_k \wedge \eta_j - s\omega_i \wedge \eta_j \wedge \eta_k\right] \\
&\quad - t^3 dt \wedge \sum_{(ijk)}\left[2\rho_k \wedge \eta_i \wedge \eta_j + \frac{1}{2}ds \wedge \eta_i \wedge \eta_j \wedge \eta_k\right] \\
&\quad + t^4 \sum_{(ijk)}\left[\eta_i \wedge (\omega_j \wedge \rho_k - \omega_k \wedge \rho_j) + \frac{1}{2}ds \wedge \omega_i \wedge \eta_j \wedge \eta_k\right] \\
&= -2t^3 \sum_{(ijk)} dt \wedge (\rho_i + 2\epsilon\omega_i) \wedge \eta_j \wedge \eta_k + t^4 d\Omega = 0, \tag{4.143}
\end{aligned}
$$

taking into account the first equality in Theorem 4.3.5, (4.130) and $s = 2\epsilon$, which hold when $d\Omega = 0$ by Theorem 4.4.2.

Hence, $dF = 0$ and the holonomy of the cone metric is contained either in $Sp(n + 1)Sp(1)$ or in $Sp(n, 1)Sp(1)$ provided $n > 1$ [157], i.e. the cone is quaternionic Kähler manifold provided $n > 1$. Note that when $n = 1$ this conclusion can not be reached in the positive definite case due to the 8-dimensional compact counter-example constructed by S. Salamon [147] (for non compact counter-examples see [10; 11]).

It is a classical result (see e.g [23] and references therein) that a quaternionic Kähler manifolds of dimension bigger than four (of arbitrary signature) are Einstein. This fact implies that the cone $N = M \times \mathbb{R}^+$ with the metric g_N must be Ricci flat (see e.g. [23]) and therefore it is locally hyperkähler since the $sp(1)$-part of the Riemannian curvature vanishes and therefore it can be trivialized locally by a parallel sections (see e.g. [23]). This means that locally there exists a $SO(3)$-matrix Ψ with smooth entries, possibly depending on t, such that the triple of two forms $(\tilde{F}_1, \tilde{F}_2, \tilde{F}_3) = \Psi \cdot (F_1, F_2, F_3)^T$ consists of closed 2-forms constitute the fundanental 2-forms of the local hyperkähler structure. Consequently $(M, \Psi \cdot \eta)$ is locally a 3-Sasakian manifold.

The fact that b) implies a) is trivial since the 4-form Ω is invariant under rotations and rescales by a constant when the metric on the horizontal space H is replaced by a homothetic to it metric.

$\square$

Remark 4.4.5. It follows from the above discussion that the cone over a $(4n + 3)$-dimensional qc manifold carries either a $Sp(n + 1)Sp(1)$ or a $Sp(n, 1)Sp(1)$ structure which is closed exactly when $d\Omega = 0$ provided the dimension is strictly bigger than seven.

Remark 4.4.6. An example of a qc structure satisfying $T^0 = U = Scal = 0$ can be obtained as follows. Let M^{4n} be a hyperkähler manifold with closed and locally exact Kähler forms $\omega_l = d\eta_l$. The total space of an $\mathbb{R}^3$-bundle over the hyperkähler manifold M^{4n} with connection 1-forms η_l is an example of a qc structure with $T^0 = U = Scal = 0$. The qc structure is determined by the three 1-forms η_l satisfying $d\eta_l = \omega_l$ which yield $T^0 = U = Scal = 0$. In particular, the quaternionic Heisenberg group which locally is the unique qc structure with flat Biquard connection on H, see [97], can be considered as an $\mathbb{R}^3$ bundle over a 4n-dimensional flat hyperkähler $\mathbb{R}^{4n}$. A compact example is provided by a T^3-bundle over a compact hyperkähler manifold M^{4n} such that each closed Kähler form ω_l represents integral cohomology classes. Indeed, since $[\omega_l]$, $1 \le l \le 3$ define integral cohomology classes on M^{4n}, the well-known result of Kobayashi [111] implies that there exists a circle bundle $S^1 \hookrightarrow M^{4n+1} \to M^{4n}$, with connection 1-form η_1 on M^{4n+1} whose curvature form is $d\eta_1 = \omega_1$. Because ω_l ($l = 2, 3$) defines an integral cohomology class on M^{4n+1}, there exists a principal circle bundle $S^1 \hookrightarrow M^{4n+2} \to M^{4n+1}$ corresponding to $[\omega_2]$ and a connection 1-form η_2 on M^{4n+2} such that $\omega_2 = d\eta_2$ is the curvature form of η_2. Using again the result of Kobayashi, one gets a T^3-bundle over M^{4n} whose total space has a qc structure satisfying $d\eta_l = \omega_l$ which yield $T^0 = U = Scal = 0$.

We do not know whether there are other examples satisfying the conditions $T^0 = U = Scal = 0$.

Chapter 5

Quaternionic contact conformal curvature tensor

5.1 Introduction

The goal of this section is to describe a tensor invariant on the tangent bundle characterizing locally the qc structures which are quaternionic contact conformally equivalent (quaternionic contact isomorphic) to the flat qc-structure on the quaternionic Heisenberg group. It was proven in [94] that the quaternionic Heisenberg group G ($\mathbb{H}$) with its "standard" left-invariant qc structure is the unique (up to a $SO(3)$-action) example of a qc structure with flat Biquard connection. We shall describe a curvature-type tensor W^{qc} defined in terms of the curvature and torsion of the Biquard connection by (5.49) involving derivatives up to second order of the horizontal metric, whose form is similar to the Weyl conformal curvature in Riemannian geometry and to the Chern-Moser invariant in CR geometry [45], see also [169]. We call W^{qc} the *quaternionic contact conformal curvature, qc conformal curvature for short*. The main result of the section is Theorem 5.3.5 in which we show that a qc structure on a $(4n+3)$-dimensional smooth manifold is locally quaternionic contact conformal to the standard flat qc structure on the quaternionic Heisenberg group G ($\mathbb{H}$) if and only if the qc conformal curvature vanishes, $W^{qc} = 0$.

In addition, we shall prove a number of useful facts. We devote a sub-section to the quaternionic Cayley transform, which turns out to be a conformal quaternionic contact automorphism between the standard 3-Sasakian structure on the quaternionic sphere S^{4n+3} minus a point and the standard qc structure on G ($\mathbb{H}$) [94]. As a consequence of Theorem 5.3.5 and the fact that the quaternionic Cayley transform is a quaternionic contact conformal equivalence between the 3-Sasakian structure on the sphere minus a point and the flat qc structure on G ($\mathbb{H}$), we obtain that a qc-manifold is locally quaternionic contact conformal to the quaternionic sphere S^{4n+3} if and only if the qc conformal curvature vanishes, $W^{qc} = 0$.

We expect that the quaternionic contact conformal curvature tensor will be a useful tool in the analysis of the quaternionic contact Yamabe problem, see Section 6. According to [168] the qc Yamabe constant of a compact qc manifold is less or equal than that of the sphere. Furthermore, if the constant is strictly less than that of the sphere the qc Yamabe problem has a solution, i.e., there is a global qc conformal transformation sending the given qc structure to a qc structure with constant qc scalar curvature. A natural conjecture is that the qc Yamabe constant of every compact locally non-flat manifold (in conformal quaternionic contact sense) is strictly less than the qc Yamabe constant of the sphere with its standard qc structure and a natural tool in investigating this conjecture is the use of the qc conformal curvature which measures the non-flatness of a qc structure.

5.2 Quaternionic contact conformal transformations

Given a qc manifold $(M, H, [g], \mathbb{Q})$ and a local qc form η such that $H = Ker\eta$ then any other qc form $\bar{\eta}$ generating the same qc structure is connected with η by $\bar{\eta} = \mu \Psi \eta$ where μ is a positive smooth function and $\Psi \in SO(3)$ is a smooth $SO(3)$ matrix. In this section we study smooth tensor invariants of a given qc structure, i.e. tensor invariants of these transformations.

A conformal quaternionic contact transformation between two quaternionic contact manifold is a diffeomorphism Φ which satisfies

$$\Phi^* \eta = \mu \, \Psi \cdot \eta$$

for some positive smooth function μ and some matrix $\Psi \in SO(3)$ with smooth functions as entries, where $\eta = (\eta_1, \eta_2, \eta_3)^t$ is considered as an element of $\mathbb{R}^3$. The Biquard connection does not change under rotations, i.e., the Biquard connection of $\Psi \cdot \eta$ and η coincides. Hence, studying qc conformal transformations we may consider only transformations

$$\Phi^* \eta = \mu \, \eta.$$

We recall the formulas for the conformal change of the corresponding Biquard connections from [94].

Let h be a positive smooth function on a qc manifold (M, η) and

$$\bar{\eta} = \frac{1}{2h} \eta$$

be a conformal deformation of the qc structure η. We will denote the objects related to $\bar{\eta}$ by over-lining the same object corresponding to η. Thus,

$$d\bar{\eta} = -\frac{1}{2h^2}\, dh \wedge \eta \,+\, \frac{1}{2h}d\eta, \qquad \bar{g} = \frac{1}{2h}g. \tag{5.1}$$

The new triple $\{\bar{\xi}_1, \bar{\xi}_2, \bar{\xi}_3\}$, determined by the conditions (4.7) defining the Reeb vector fields, is

$$\bar{\xi}_s \;=\; 2h\,\xi_s \,+\, I_s\nabla h, \tag{5.2}$$

where ∇h is the *horizontal gradient* defined by

$$\nabla h = dh(e_a)e_a \tag{5.3}$$

using the fixed orthonormal basis $e_1, \ldots, e_{4n}$ of the horizontal space H equipped with the metric determined by η, see (4.1). The *horizontal sub-Laplacian* and the norm of the horizontal gradient are defined respectively by

$$\triangle h \;=\; tr_H^g(\nabla dh) \;=\; \nabla dh(e_\alpha, e_\alpha), \qquad |\nabla h|^2 \;=\; dh(e_\alpha)\,dh(e_\alpha). \tag{5.4}$$

The Biquard connections ∇ and $\bar{\nabla}$ are connected by a (1,2) tensor P,

$$\bar{\nabla}_A B = \nabla_A B + P(A, B), \qquad A, B \in \Gamma(TM). \tag{5.5}$$

Condition (4.11) yields

$$g(P(X,Y), Z) - g(P(Y,X), Z) = -h^{-1}\sum_{s=1}^{3}\omega_s(X,Y)dh(I_s Z), \tag{5.6}$$

while $\bar{\nabla}\bar{g} = 0$ implies

$$g(P(A,Y), Z) + g(P(A,Z), Y) = -h^{-1}dh(A)g(Y,Z). \tag{5.7}$$

Set $A = X$ into (5.7) and combine the obtained equation with (5.6) to determine the tensor P

$$2hg(P(X,Y),Z) = -dh(X)g(Y,Z) + \sum_{s=1}^{3} dh(I_s X)\omega_s(Y,Z) - dh(Y)g(Z,X)$$

$$- \sum_{s=1}^{3} dh(I_s Y)\omega_s(Z,X) + dh(Z)g(X,Y) - \sum_{s=1}^{3} dh(I_s Z)\omega_s(X,Y). \tag{5.8}$$

We calculate

$$2h\bar{T}(\bar{\xi}_i, X, Y) - 2hT(\xi_i, X, Y) - g(P(\bar{\xi}_i, X), Y)$$
$$= -g(P(X, \bar{\xi}_i), Y) = -\bar{\alpha}_j(X)dh(I_k Y) + \bar{\alpha}_k(X)dh(I_j Y)$$
$$+ \alpha_j(X)dh(I_k Y) - \alpha_k(X)dh(I_j Y) - \nabla dh(X, I_i Y)$$
$$= -\nabla dh(X, I_i Y) + h^{-1}\Big[dh(I_k X)dh(I_j Y) - dh(I_j X)dh(I_k Y)\Big], \tag{5.9}$$

where we applied (4.10), (4.16), (5.1)and (5.2) several times.

When we take the symmetric part of (5.9), then apply (4.20) and (5.7) with $A = \bar{\xi}_i$ gives

$$4h\bar{T}^0(\bar{\xi}_i, X, Y) - 4hT^0(\xi_i, X, Y) + h^{-1}dh(\bar{\xi}_i)g(X, Y)$$
$$= -\nabla dh(X, I_iY) - \nabla dh(Y, I_iX).$$

The last equality together with (4.46) and (5.2) yield

$$\bar{T}^0(X, Y) - T^0(X, Y) + \frac{1}{2h} \sum_{s=1}^{3} dh(\xi_s)\omega_s(X, Y)$$
$$= \frac{1}{4h} \left(3\nabla dh(Y, X) - \sum_{s=1}^{3} \nabla dh(I_sX, I_sY) \right). \quad (5.10)$$

The identity $d^2 = 0$ implies $\nabla dh(X, Y) - \nabla dh(Y, X) = -dh(T(X, Y))$. Applying (4.11), we have

$$\nabla dh(X, Y) = [\nabla dh]_{[sym]}(X, Y) - \sum_{s=1}^{3} dh(\xi_s)\omega_s(X, Y), \quad (5.11)$$

where $[.]_{[sym]}$ denotes the symmetric part of the corresponding (0,2)-tensor.

A substitution of (5.11) in (5.10) gives the important identity [94]

$$\bar{T}^0(X, Y) = T^0(X, Y) + h^{-1}[\nabla dh]_{[sym][-1]}(X, Y), \quad (5.12)$$

where $[\nabla dh]_{[sym][-1]}$ denotes the symmetric [-1] part of ∇dh given by

$$[\nabla dh]_{[sym][-1]}(X, Y)$$
$$= \frac{1}{4} \left(3\nabla dh(X, Y) - \sum_{s=1}^{3} \nabla dh(I_sX, I_sY) + 4\sum_{s=1}^{3} dh(\xi_s)\omega_s(X, Y) \right).$$
$$(5.13)$$

Taking the skew-symmetric part of (5.9), then applying the first part of (4.13) and (5.7) with $A = \bar{\xi}_i$ we obtain

$$4h\bar{S}(\bar{\xi}_i, X, Y) - 4hS(\xi_i, X, Y)$$
$$= 2g(P(\bar{\xi}_i, X), Y) + 2dh(\xi_i)g(X, Y) - \nabla dh(X, I_iY) + \nabla dh(Y, I_iX)$$
$$+ 2h^{-1}\Big[dh(I_kX)dh(I_jY) - dh(I_jX)dh(I_kY)\Big]. \quad (5.14)$$

The second equation in (4.46) together with Lemma 4.2.7 and (5.11) applied to (5.14) yield

$$4h\bar{U}(X, Y) - 4hU(X, Y) = -2g(P(\bar{\xi}_i, I_iX), Y)$$
$$+ \nabla dh(X, Y) + \nabla dh(I_iX, I_iY) + 2dh(\xi_j)\omega_j(X, Y) + 2dh(\xi_k)\omega_k(X, Y)$$
$$- 2h^{-1}\Big[dh(I_jX)dh(I_jY) + dh(I_kX)dh(I_kY)\Big]. \quad (5.15)$$

$$4h\bar{U}(I_j X, I_j Y) - 4hU(I_j X, I_j Y) = -2g(P(\bar{\xi}_i, I_k X), I_j Y)$$
$$+ \nabla dh(I_j X, I_j Y) + \nabla dh(I_k X, I_k Y) + 2dh(\xi_j)\omega_j(X, Y) - 2dh(\xi_k)\omega_k(X, Y)$$
$$- 2h^{-1}\Big[dh(X)dh(Y) + dh(I_i X)dh(I_i Y)\Big]. \quad (5.16)$$

Since both connections are quaternionic connections, we obtain from (4.10) consequently that

$$\Big[-\bar{\alpha}_j(\bar{\xi}_i) + \alpha_j(\bar{\xi}_i)\Big]\omega_k(X, Y) - \Big[-\bar{\alpha}_k(\bar{\xi}_i) + \alpha_k(\bar{\xi}_i)\Big]\omega_j(X, Y)$$
$$= g(\bar{\nabla}_{\xi_i} I_i)X, Y) - g(\nabla_{\xi_i} I_i)X, Y)$$
$$= g(P(\bar{\xi}_i, I_i X), Y) - g(P(\bar{\xi}_i, X), I_i Y); \quad (5.17)$$

$$\Big[-\bar{\alpha}_i(\bar{\xi}_i) + \alpha_i(\bar{\xi}_i)\Big]g(X, Y) - \Big[-\bar{\alpha}_j(\bar{\xi}_i) + \alpha_j(\bar{\xi}_i)\Big]\omega_k(X, Y)$$
$$= g(\bar{\nabla}_{\xi_i} I_k)X, I_j Y) - g(\nabla_{\xi_i} I_k)X, I_j Y)$$
$$= g(P(\bar{\xi}_i, I_k X), I_j Y) + g(P(\bar{\xi}_i, X), I_i Y). \quad (5.18)$$

We calculate from (4.92) and (4.93) applying (5.1) and (5.2) that

$$-\bar{\alpha}_j(\bar{\xi}_i) + \alpha_j(\bar{\xi}_i) = 2dh(\xi_k); \quad -\bar{\alpha}_k(\bar{\xi}_i) + \alpha_k(\bar{\xi}_i) = -2dh(\xi_j);$$
$$-\bar{\alpha}_i(\bar{\xi}_i) + \alpha_i(\bar{\xi}_i) = \frac{1}{16n(n+2)}\left(\overline{Scal} - 2hScal\right) + \frac{1}{2h}|\nabla h|^2. \quad (5.19)$$

Summing up (5.17) and (5.18), insert the result into the sum of (5.15) and (5.16) using Lemma 4.2.7 and (5.19), we obtain

$$\bar{U}(X, Y) = U(X, Y) + \frac{1}{2h}\left[\nabla dh - \frac{2}{h}dh \otimes dh\right]_{[3]}(X, Y)$$
$$- \frac{1}{4h}\Big[-\bar{\alpha}_i(\bar{\xi}_i) + \alpha_i(\bar{\xi}_i)\Big]g(X, Y), \quad (5.20)$$

where $\left[\nabla dh - \frac{2}{h}dh \otimes dh\right]_{[3]}$ is the [3]-component given by

$$\left[\nabla dh - \frac{2}{h}dh \otimes dh\right]_{[3]}(X, Y) = \frac{1}{4}\nabla dh(X, Y) + \frac{1}{4}\sum_{s=1}^{3}\nabla dh(I_s X, I_s Y)$$
$$- \frac{1}{2h}\left(dh(X)dh(Y) + \sum_{s=1}^{3}dh(I_s X)dh(I_s Y)\right). \quad (5.21)$$

The trace of (5.20) with the help of (5.21) yields

$$\Big[-\bar{\alpha}_i(\bar{\xi}_i) + \alpha_i(\bar{\xi}_i)\Big] = \frac{1}{2n}\Delta h - \frac{1}{nh}|\nabla h|^2, \quad (5.22)$$

since the tensors $\bar{U}$ and U are completely trace-free.

Substitute (5.22) into (5.20) to obtain [94]

$$\bar{U}(X,Y) = U(X,Y) + \frac{1}{2h}\left[\nabla dh - \frac{2}{h}dh \otimes dh\right]_{[3][0]}(X,Y), \qquad (5.23)$$

where $\left[\nabla dh - \frac{2}{h}dh \otimes dh\right]_{[3][0]}$ denotes the trace-free part of the [3]-component,

$$\left[\nabla dh - \frac{2}{h}dh \otimes dh\right]_{[3][0]} = \left[\nabla dh - \frac{2}{h}dh \otimes dh\right]_{[3]} - \frac{1}{4n}\left(\Delta h - \frac{2}{h}|\nabla h|^2\right)g.$$

Now, (5.15) and (5.23) yield the next transformation formula [94]

$$\begin{aligned}
g(P_{\bar{\xi}_i}X,Y) &= \frac{1}{4}\left[\nabla dh(X, I_iY) - \nabla dh(I_iX, Y)\right] \\
&+ \frac{1}{4}\left[\nabla dh(I_jX, I_kY) - \nabla dh(I_kX, I_jY)\right] \\
&- \frac{1}{2h}\left[dh(I_kX)dh(I_jY) - dh(I_jX)dh(I_kY)\right] \\
&- \frac{1}{2h}\left[dh(I_iX)dh(Y) - dh(X)dh(I_iY)\right] \\
&+ \frac{1}{4n}\left(-\Delta h + \frac{2}{h}|\nabla h|^2\right)\omega_i(X,Y) \\
&- dh(\xi_k)\omega_j(X,Y) + dh(\xi_j)\omega_k(X,Y). \quad (5.24)
\end{aligned}$$

Compare (5.22) with the second line of (5.19) to get the formula connecting the qc scalar curvatures of $\bar{\eta} = \frac{1}{2h}\eta$ and η [24] (see also [94])

$$\overline{Scal} = 2hScal + 8(n+2)\Delta h - \frac{1}{h}8(n+2)^2|\nabla h|^2. \qquad (5.25)$$

Theorem 4.4.2 together with (5.12) and (5.23) lead to the next result established in [94]

Corollary 5.2.1. *The qc Einstein condition is preserved under a qc conformal transformation $\bar{\eta} = \frac{1}{2h}\eta$ if and only if the function h satisfies the following two PDEs*

$$3\nabla dh(X,Y) - \sum_{s=1}^{3}\nabla dh(I_sX, I_sY) + 4\sum_{s=1}^{3}dh(\xi_s)\omega_s(X,Y) = 0; \qquad (5.26)$$

$$\begin{aligned}
&\nabla dh(X,Y) + \sum_{s=1}^{3}\nabla dh(I_sX, I_sY) - \frac{2}{h}dh(X)dh(Y) \\
&- \frac{2}{h}\sum_{s=1}^{3}dh(I_sX)dh(I_sY)\right) - \frac{1}{4n}\left(\Delta h - \frac{2}{h}|\nabla h|^2\right)g(X,Y) = 0. \quad (5.27)
\end{aligned}$$

5.2.1 *The quaternionic Cayley transform*

Next we bring into consideration the quaternionic Heisenberg group $G\,(\mathbb{H})$ and the 3-Sasakian sphere S both of dimension $4n + 3$. In Section 2.3.1 equation (2.30) we defined a Cayley transform on all groups of Heisenberg type. Here we focus on the quaternionic Cayley transform which turns out to be a qc conformal transformation between the standard flat qc structure on $G\,(\mathbb{H})$ and the standard 3-Sasakian structure on the sphere minus a point [94].

Let us identify $G\,(\mathbb{H})$ with the boundary Σ of a Siegel domain in $\mathbb{H}^n \times \mathbb{H}$, see also (2.19),

$$\Sigma = \{(q', p') \in \mathbb{H}^n \times \mathbb{H} : \Re\, p' = |q'|^2\},$$

by using the map

$$(q', \omega') \mapsto (q', |q'|^2 - \omega').$$

The standard contact form $\tilde{\Theta}$, written as a purely imaginary quaternion valued form, is given by (4.127). Since $dp' = q' \cdot d\bar{q}' + dq' \cdot \bar{q}' - d\omega'$, under the identification of $G\,(\mathbb{H})$ with Σ we also have

$$\tilde{\Theta} = -\frac{1}{2}dp' + dq' \cdot \bar{q}'.$$

Taking into account that $\tilde{\Theta}$ is purely imaginary, the last equation can be written also in the following form

$$\tilde{\Theta} = \frac{1}{4}(d\bar{p}' - dp') + \frac{1}{2}dq' \cdot \bar{q}' - \frac{1}{2}q' \cdot d\bar{q}'.$$

The Cayley transform is the map from the sphere $S = \{|q|^2 + |p|^2 = 1\} \subset \mathbb{H}^n \times \mathbb{H}$ minus a point to the quaternionic Heisenberg group Σ defined by $\mathcal{C} : S \mapsto \Sigma$, $(q', p') = \mathcal{C}\left((q, p)\right)$, where

$$q' = (1 + p)^{-1}\, q, \qquad p' = (1 + p)^{-1}\,(1 - p).$$

The inverse map is $(q, p) = \mathcal{C}^{-1}\left((q', p')\right)$ given by

$$q = 2(1 + p')^{-1}\, q', \qquad p = (1 + p')^{-1}\,(1 - p').$$

The Cayley transform maps S minus a point to Σ since

$$\Re\, p' = \Re\frac{(1 + \bar{p})(1 - p)}{|1 + p|^2} = \Re\frac{1 - |p|}{|1 + p|^2} = \frac{|q|^2}{|1 + p|^2} = |q'|^2.$$

Writing the Cayley transform in the form $(1+p)q' = q$, $(1+p)p' = 1-p$, gives $dp \cdot q' + (1+p) \cdot dq' = dq$, $dp \cdot p' + (1+p) \cdot dp' = -dp$, from where we find

$$
\begin{aligned}
dp' &= -2(1+p)^{-1} \cdot dp \cdot (1+p)^{-1} \\
dq' &= (1+p)^{-1} \cdot [dq - dp \cdot (1+p)^{-1} \cdot q].
\end{aligned} \tag{5.28}
$$

The Cayley transform is a conformal quaternionic contact diffeomorphism between the quaternionic Heisenberg group with its standard quaternionic contact structure $\tilde{\Theta}$ and the sphere minus a point with its standard qc structure $\tilde{\eta}$, a fact which can be seen as follows. Equations (5.28) imply the following identities

$$
\begin{aligned}
2\mathcal{C}^* \tilde{\Theta} &= -(1+\bar{p})^{-1} \cdot d\bar{p} \cdot (1+\bar{p})^{-1} + (1+p)^{-1} \cdot dp \cdot (1+p)^{-1} \\
&\quad + (1+p)^{-1} \left[dq - dp \cdot (1+p)^{-1} \cdot q \right] \cdot \bar{q} \cdot (1+\bar{p})^{-1} \\
&\quad - (1+p)^{-1} q \cdot \left[d\bar{q} - \bar{q} \cdot (1+\bar{p})^{-1} \cdot d\bar{p} \right] \cdot (1+\bar{p})^{-1} \\
&= (1+p)^{-1} \left[dp \cdot (1+p)^{-1} \cdot (1+\bar{p}) - |q|^2 dp \cdot (1+p)^{-1} \right] (1+\bar{p})^{-1} \\
&\quad + (1+p)^{-1} \left[-(1+p) \cdot (1+\bar{p})^{-1} \cdot d\bar{p} + |q|^2 (1+p)^{-1} d\bar{p} \right] (1+\bar{p})^{-1} \\
&\quad + (1+p)^{-1} \left[dq \cdot \bar{q} - q \cdot d\bar{q} \right] (1+\bar{p})^{-1} = |1+p|^{-2} \lambda \tilde{\eta} \bar{\lambda}, \quad (5.29)
\end{aligned}
$$

where $\lambda = |1+p| (1+p)^{-1}$ is a unit quaternion and $\tilde{\eta}$ is the standard quaternionic contact form on the sphere,

$$
\tilde{\eta} = dq \cdot \bar{q} + dp \cdot \bar{p} - q \cdot d\bar{q} - p \cdot d\bar{p}. \tag{5.30}
$$

Since $|1+p| = 2|1+p'|^{-1}$ we have $\lambda = (1+p')|1+p'|^{-1}$ and the equation (5.29) can be put in the form

$$
\lambda \cdot (\mathcal{C}^{-1})^* \tilde{\eta} \cdot \bar{\lambda} = \frac{8}{|1+p'|^2} \tilde{\Theta}.
$$

Remark 5.2.2. Notice that the standard qc contact form we consider here is twice the 3-Sasakian form on S^{4n+3}, which has qc scalar curvature equal to $16n(n+2)$ by Theorem 4.4.3. Thus, the qc scalar curvature $\tilde{S} = \widetilde{Scal}$ of the standard qc structure (5.30) on S^{4n+3} is

$$
\tilde{S} = \frac{1}{2}(Q+2)(Q-6) = 8n(n+2), \tag{5.31}
$$

where, as usual, $Q = 4n + 6$ is the homogeneous dimension.

5.3 qc conformal curvature

We calculate the relation between the curvature tensors $\bar{R}$ and R of the Biaquard connections $\bar{\nabla}$ and ∇ of two qc conformally equivalent qc structures $\bar{\eta} = \frac{1}{2h}\eta$.

We use the notation for the Kulkarni-Nomizu product of two (not necessarily symmetric) tensors of type (0,2), for example, if L is a (0,2) tensor, we have

$$(\omega_s \otimes L)(X,Y,Z,V) := \omega_s(X,Z)L(Y,V) + \omega_s(Y,V)L(X,Z)$$
$$- \omega_s(Y,Z)L(X,V) - \omega_s(X,V)L(Y,Z).$$

We also use the usual conventions

$$I_s L\,(X,Y) = g(I_s LX, Y) = -L(X, I_s Y).$$

With a standard computation based on (5.5), (5.8), (5.24) the curvature tensors $\bar{R}$ and R are connected by [97]

$$2hg(\bar{R}(X,Y)Z,V) - g(R(X,Y)Z,V) = -(g \otimes M)(X,Y,Z,V)$$

$$-\sum_{s=1}^{3}(\omega_s \otimes I_s M)(X,Y,Z,V) + \frac{1}{2}\sum_{(i,j,k)}\omega_i(X,Y)\Big[M(Z,I_iV) - M(I_iZ,V)\Big]$$

$$+ \frac{1}{2}\sum_{(i,j,k)}\omega_i(X,Y)\Big[M(I_jZ,I_kV) - M(I_kZ,I_jV)\Big]$$

$$- g(Z,V)\Big[M(X,Y) - M(Y,X)\Big] + \sum_{s=1}^{3}\omega_s(Z,V)\Big[M(X,I_sY) - M(Y,I_sX)\Big]$$

$$+ \frac{1}{2n}\sum_{(i,j,k)}M_i\Big[\omega_j(X,Y)\omega_k(Z,V) - \omega_k(X,Y)\omega_j(Z,V)\Big]$$

$$- \frac{1}{2n}(trM)\sum_{s=1}^{3}\omega_s(X,Y)\omega_s(Z,V), \quad (5.32)$$

where $\sum_{(i,j,k)}$ denotes the cyclic sum, the (0,2) tensor M is given by

$$M(X,Y) = \frac{1}{2h}\nabla dh(X,Y)$$

$$- \frac{1}{4h^2}\Big[dh(X)dh(Y) + \sum_{s=1}^{3}dh(I_sX)dh(I_sY) + \frac{1}{2}g(X,Y)|dh|^2\Big] \quad (5.33)$$

and $trM = M(e_a, e_a), M_s = M(e_a, I_s e_a)$ are its traces.

Using (5.33) and (5.11), we obtain

$$trM = \frac{1}{2h}\left(\triangle h - \frac{n+2}{h}|dh|^2\right),$$

$$M_s = \frac{1}{4h}\left(\nabla dh(e_a, I_s e_a) - \nabla dh(I_s e_a, e_a)\right) = -\frac{2n}{h}dh(\xi_s).$$
(5.34)

Suitable traces of (5.32) imply

$$\overline{\rho}_i(X, I_i Y) - \rho_i(X, I_i Y) = -M(X, Y) - M(I_i Y, I_i X)$$

$$- \frac{trM}{2n}g(X, Y) + \frac{M_j}{2n}\omega_j(X, Y) + \frac{M_k}{2n}\omega_k(X, Y)$$

$$= -M(X, Y) - M(I_i X, I_i Y)) - \frac{trM}{2n}g(X, Y) - \frac{1}{h}dh(\xi_i)\omega_i(X, Y), \quad (5.35)$$

where we use (5.33), (5.34) and (5.11) to establish the third line.

The equations (5.35) yield

$$\sum_{s=1}^{3}\overline{\rho}_s(X, I_s Y) - \sum_{s=1}^{3}\rho_s(X, I_s Y) = -2M(X, Y)$$

$$- 4M_{[3]}(X, Y) - \frac{3trM}{2n}g(X, Y) - \frac{1}{h}\sum_{s=1}^{3}dh(\xi_s)\omega_s(X, Y)$$

$$= -2M_{[sym]}(X, Y) - 4M_{[3]}(X, Y) - \frac{3trM}{2n}g(X, Y), \quad (5.36)$$

where we apply (5.11) to obtain the last line.

Using (5.33) and the fact that the [3]-component $(\nabla dh)_{[3]}$ of ∇dh on H is symmetric, we obtain that the [3]-component of M is symmetric, $M_{[3]} = M_{[3][sym]} = M_{[sym][3]}$.

The two $Sp(n)Sp(1)$-invariant components of (5.36) are given by

$$\left(\sum_{s=1}^{3}\overline{\rho}_s(X, I_s Y)\right)_{[-1]} - \left(\sum_{s=1}^{3}\rho_s(X, I_s Y)\right)_{[-1]}$$

$$= -2M_{[sym][-1]}(X, Y). \quad (5.37)$$

$$\left(\sum_{s=1}^{3}\overline{\rho}_s(X, I_s Y)\right)_{[3]} - \left(\sum_{s=1}^{3}\rho_s(X, I_s Y)\right)_{[3]}$$

$$= -6M_{[sym][3]}(X, Y) - \frac{3trM}{2n}g(X, Y). \quad (5.38)$$

On the other hand (4.69) implies

$$\sum_{s=1}^{3} \rho_s(X, I_s Y)$$

$$= -\frac{3}{8n(n+2)} Scal.g(X,Y) - T^0(X,Y) - 6U(X,Y). \quad (5.39)$$

Taking the trace in (5.38) and applying (5.39), we obtain

$$\frac{\overline{Scal}}{2h} - Scal = 8(n+2)trM. \quad (5.40)$$

Now, (5.40) yields

$$trM.g(X,Y) = \frac{\overline{Scal}}{8n(n+2)}\overline{g}(X,Y) - \frac{Scal}{8n(n+2)}g(X,Y). \quad (5.41)$$

We calculate from (5.38), (5.39) and (5.41) that

$$M_{[sym][3]}(X,Y) = \frac{1}{6}\left(\sum_{s=1}^{3} \rho_s(X, I_s Y)\right)_{[3]} + \frac{Scal}{32n(n+2)}g(X,Y)$$

$$- \frac{1}{6}\left(\sum_{s=1}^{3} \overline{\rho}_s(X, I_s Y)\right)_{[3]} - \frac{\overline{Scal}}{32n(n+2)}\overline{g}(X,Y)$$

$$= \left[\overline{U}(X,Y) + \frac{\overline{Scal}}{32n(n+2)}\overline{g}(X,Y)\right] - \left[U(X,Y) + \frac{Scal}{32n(n+2)}g(X,Y)\right]. \quad (5.42)$$

Similarly, (5.37) together with (5.39) imply

$$M_{[sym][-1]}(X,Y) = \frac{1}{2}\overline{T}^0(X,Y) - \frac{1}{2}T^0(X,Y). \quad (5.43)$$

We get from (5.42) and (5.43) that

$$M_{[sym]}(X,Y) = \overline{L}(X,Y) - L(X,Y), \quad (5.44)$$

where the symmetric (0,2) tensor L is given by

$$L(X,Y) = \frac{1}{2}T^0(X,Y) + U(X,Y) + \frac{Scal}{32n(n+2)}g(X,Y). \quad (5.45)$$

We obtain from (5.33) and (5.11) that

$$M(X,Y) = M_{[sym]}(X,Y) - \frac{1}{2h}\sum_{s=1}^{3} dh(\xi_s)\omega_s(X,Y). \quad (5.46)$$

Substitute (5.45) into (5.44) and put the result into (5.46), insert the obtained equality in (5.32) and use (5.34) to obtain [97]

$$2h\overline{WR}(X,Y,Z,V) = WR(X,Y,Z,V), \tag{5.47}$$

where the tensor WR is given by

$$
\begin{aligned}
WR(X,Y,Z,V) &= R(X,Y,Z,V) \\
&+ (g \oslash L)(X,Y,Z,V) + \sum_{s=1}^{3}(\omega_s \oslash I_s L)(X,Y,Z,V) \\
&- \frac{1}{2}\sum_{(i,j,k)}\omega_i(X,Y)\Big[L(Z,I_iV) - L(I_iZ,V) + L(I_jZ,I_kV) - L(I_kZ,I_jV)\Big] \\
&- \sum_{s=1}^{3}\omega_s(Z,V)\Big[L(X,I_sY) - L(I_sX,Y)\Big] + \frac{1}{2n}(trL)\sum_{s=1}^{3}\omega_s(X,Y)\omega_s(Z,V).
\end{aligned}
\tag{5.48}
$$

Recalling the definitions (4.4) of the two $Sp(n)Sp(1)$-invariant parts of an endomorphism of H, comparing (5.48) with (4.104) and applying (5.45), we obtain the next Proposition.

Proposition 5.3.1. *On a QC manifold the [−1]-part with respect to the first two arguments of the tensor WR vanishes identically,*

$$
\begin{aligned}
WR_{[-1]}(X,Y,Z,V) \\
= \frac{1}{4}\Big[3WR(X,Y,Z,V) - \sum_{s=1}^{3}WR(I_sX,I_sY,Z,V)\Big] = 0.
\end{aligned}
$$

The [3]-part with respect to the first two arguments of the tensor WR

coincides with WR and has the expression

$$WR(X,Y,Z,V) = WR_{[3]}(X,Y,Z,V)$$

$$= \frac{1}{4}\Big[WR(X,Y,Z,V) + \sum_{s=1}^{3} WR(I_s X, I_s Y, Z, V)\Big]$$

$$= \frac{1}{4}\Big[R(X,Y,Z,V) + \sum_{s=1}^{3} R(I_s X, I_s Y, Z, V)\Big]$$

$$- \frac{1}{2}\sum_{s=1}^{3} \omega_s(Z,V)\Big[T^0(X, I_s Y) - T^0(I_s X, Y)\Big]$$

$$+ \frac{Scal}{32n(n+2)}\Big[(g \oslash g)(X,Y,Z,V) + \sum_{s=1}^{3}(\omega_s \oslash \omega_s)(X,Y,Z,V)\Big]$$

$$+ (g \oslash U)(X,Y,Z,V) + \sum_{s=1}^{3}(\omega_s \oslash I_s U)(X,Y,Z,V). \quad (5.49)$$

Definition 5.3.2. We denote with W^{qc} the tensor WR (5.48) considered as a tensor of type (1,3) with respect to the horizontal metric on H, $g(W^{qc}(X,Y)Z,V) = WR(X,Y,Z,V)$ and call it *the quaternionic contact conformal curvature.*

In view of (5.47) we obtain the next important result discovered in [97]

Theorem 5.3.3. *The tensor WR is covariant while the tensor W^{qc} is invariant under qc conformal transformations, i.e. if*

$$\bar{\eta} = (2h)^{-1}\Psi\eta \quad then \quad 2hWR_{\bar{\eta}} = WR_{\eta}, \quad W^{qc}_{\bar{\eta}} = W^{qc}_{\eta},$$

for any smooth positive function h and any $SO(3)$-matrix Ψ.

The properties of the qc conformal curvature established in [97] we described in the following

Proposition 5.3.4. *The tensor WR is completely trace-free, i.e.*

$$Ric(WR) = \rho_s(WR) = \tau_s(WR) = \zeta_s(WR) = 0.$$

Proof. Lemma 4.2.6, Lemma (4.2.7) and (5.45) imply the following identities

$$T^0(\xi_s, I_s X, Y) = \frac{1}{2}\Big[L(X,Y) - L(I_s X, I_s Y)\Big], \quad (5.50)$$

$$U(X,Y) = -\frac{1}{4n} tr \, L \, g(X,Y)$$
$$+ \frac{1}{4}\Big[L(X,Y) + L(I_1X, I_1Y) + L(I_2X, I_2Y) + L(I_3X, I_3Y)\Big], \quad (5.51)$$

$$T(\xi_i, X, Y) = -\frac{1}{2}\Big[L(I_iX, Y) + L(X, I_iY)\Big] + U(I_iX, Y)$$
$$= -\frac{1}{4}L(I_iX, Y) - \frac{3}{4}L(X, I_iY) - \frac{1}{4}L(I_kX, I_jY)$$
$$+ \frac{1}{4}L(I_jX, I_kY) - \frac{1}{4n}(tr \, L) \, g(I_iX, Y). \quad (5.52)$$

After a substitution of (5.50) and (5.51) in the first four equations of Theorem 4.3.5 we derive

$$Ric(X,Y) = \frac{2n+3}{2n} tr \, L \, g(X,Y) + \frac{8n+11}{2} L(X,Y)$$
$$+ \frac{3}{2}\Big[L(I_iX, I_iY) + L(I_jX, I_jY) + L(I_kX, I_kY)\Big], \quad (5.53)$$

$$\rho_i(X,Y) = L(X, I_iY) - L(I_iX, Y) - \frac{1}{2n} tr L \, \omega_i(X,Y), \quad (5.54)$$

$$\tau_i(X,Y) = -\frac{1}{n} tr \, L \, \omega_i(X,Y)$$
$$- \frac{n+2}{2n}\Big[L(I_iX, Y) - L(X, I_iY) + L(I_kX, I_jY) - L(I_jX, I_kY)\Big] \quad (5.55)$$

$$\zeta_i(X,Y) = \frac{2n-1}{8n^2} tr \, L \, \omega_i(X,Y) + \frac{3}{8n}L(I_iX, Y)$$
$$- \frac{8n+3}{8n}L(X, I_iY) + \frac{1}{8n}\Big[L(I_kX, I_jY) - L(I_jX, I_kY)\Big]. \quad (5.56)$$

Taking the corresponding traces in (5.48), using also (5.53)-(5.56), it is straightforward to verify the claim. □

It turns out that the qc conformal curvature is the only local obstruction for a qc conformal flatness of a given qc structure.

Theorem 5.3.5. [97] *A qc structure on a (4n + 3)-dimensional smooth manifold is locally quaternionic contact conformal to the standard flat qc structure on the quaternionic Heisenberg group $G\,(\mathbb{H})$ if and only if the qc conformal curvature vanishes, $W^{qc} = 0$.*

Proof. Here we only sketch the proof and refer to [97] for a complete proof of this theorem. In Chapter 7 we give a complete new proof of the Cartan-Chern-Moser theorem in the CR case illustrating our approach in a somewhat simpler setting.

Suppose $W^{qc} = 0$, hence $WR = 0$. In order to prove Theorem 5.3.5 we search for a conformal factor such that after a conformal transformation using this factor the new qc structure has Biquard connection which is flat when restricted to the common horizontal space H. After we achieve this task we can invoke Proposition 4.3.15 and conclude that the given structure is locally qc conformal to the flat qc structure on the quaternionic Heisenberg group $G(\mathbb{H})$. With this considerations in mind, it is then sufficient to find (locally) a solution h of equation (5.46) with $M_{[sym]} = -L$. In fact, a substitution of (5.46) in (5.32) and an application of the condition $W^{qc} = 0 = WR$ allows us to see that the qc structure $\bar{\eta} = \frac{1}{2h}\eta$ has flat Biquard connection.

Let us consider the following overdetermined system of partial differential equations with respect to an unknown function u

$$\nabla du(X, Y) = -du(X)du(Y)$$

$$+ \sum_{s=1}^{3} \left[du(I_s X)du(I_s Y) - du(\xi_s)\omega_s(X, Y) \right] + \frac{1}{2}g(X, Y)|\nabla u|^2 - L(X, Y),$$

$$(5.57)$$

$$\nabla du(X, \xi_i) = \mathbb{B}(X, \xi_i) - L(X, I_i du) + \frac{1}{2}du(I_i X)|\nabla u|^2$$

$$- du(X)du(\xi_i) - du(I_j X)du(\xi_k) + du(I_k X)du(\xi_j), \quad (5.58)$$

$$\nabla du(\xi_i, \xi_i) = -\mathbb{B}(\xi_i, \xi_i) + \mathbb{B}(I_i du, \xi_i)$$

$$+ \frac{1}{4}|\nabla u|^4 - (du(\xi_i))^2 + (du(\xi_j))^2 + (du(\xi_k))^2, \quad (5.59)$$

$$\nabla du(\xi_j, \xi_i) = -\mathbb{B}(\xi_j, \xi_i) + \mathbb{B}(I_i du, \xi_j)$$

$$- 2du(\xi_i)du(\xi_j) - \frac{Scal}{16n(n+2)}du(\xi_k), \quad (5.60)$$

$$\nabla du(\xi_k, \xi_i) = -\mathbb{B}(\xi_k, \xi_i) + \mathbb{B}(I_i du, \xi_k)$$

$$- 2du(\xi_i)du(\xi_k) + \frac{Scal}{16n(n+2)}du(\xi_j). \quad (5.61)$$

Here the tensor L is given by (5.45). The tensors $\mathbb{B}(X, \xi_i)$ and $\mathbb{B}(\xi_i, \xi_j)$ do not depend on the unknown function u and are defined in terms of L and its first and second horizontal (covariant) derivatives as follows

$$\mathbb{B}(X, \xi_i) = \frac{1}{2(2n+1)}(\nabla_{e_a} L)(I_i e_a, X)$$
$$+ \frac{1}{6(2n+1)}\Big((\nabla_{e_a} L)(e_a, I_i X) - \nabla_{I_i X} tr\, L\Big),$$
$$\mathbb{B}(\xi_s, \xi_t) = \frac{1}{4n}\Big[(\nabla_{e_a}\mathbb{B})(I_s e_a, \xi_t) + L(e_u, e_b)I_i(I_t e_a, I_s e_b)\Big].$$

If we make the substitution

$$2u = \ln h, \qquad 2hdu = dh, \qquad \nabla dh = 2h\nabla du + 4hdu \otimes du,$$

in (5.33) we recognize that (5.46) transforms into (5.57). Therefore, our goal is to show that equation (5.57) has a solution, for which it is sufficient to verify the Ricci identities (see below). However, if equation (5.57) has a smooth solution then (5.58)-(5.61) appear as necessary conditions, so we considered the complete system (5.57)-(5.61) and reduced the question to showing that this system has (locally) a smooth solution.

The integrability conditions for the above considered over-determined system are furnished by the Ricci identity

$$\nabla^2 du(A, B, C) - \nabla^2 du(B, A, C)$$
$$= -R(A, B, C, du) - \nabla du((T(A, B), C). \quad (5.62)$$

The proof of Theorem 5.3.5 could be achieved by considering all possible cases of (5.62) and showing that the vanishing of the quaternionic contact conformal curvature tensor W^{qc} implies (5.62), which guaranties the existence of a local smooth solution to the system (5.57)-(5.61). We refer to [97] for details.   □

We end this section with a few remarks.

First, we note that for locally 3-Sasakian manifolds a curvature invariant under very special quaternionic contact conformal transformations, which preserve the 3-Sasakian condition, is defined in [4]. It is shown that the vanishing of this invariant is equivalent to the structure being locally isometric to the 3-Sasaki structure on the sphere. In particular, this shows that the standard 3-Sasakian structure on the sphere is locally rigid with respect to qc conformal transformations preserving the 3-Sasakian condition.

The second remark concerns an alternative approach to conformal flatness. Following the work of Cartan and Tanaka, a qc structure can be considered as an example of what has become known as a parabolic geometry. The quaternionic Heisenberg group, as well as, the quaternionic sphere, due to the Cayley transform, provide the flat models of such a geometry. It is well known that the curvature of the corresponding regular Cartan connection is the obstruction for the local flatness. However, the Cartan curvature is not a tensor field on the tangent bundle and it is highly nontrivial to extract a tensor field involving the lowest order derivatives of the structure which implies the vanishing of the obstruction. Theorem 5.3.5 suggests that a necessary and sufficient condition for the vanishing of the Cartan curvature of a qc structure is the vanishing of the qc conformal curvature tensor, $W^{qc} = 0$. This was confirmed in [9] employing results developed in [94] and the general theory of parabolic geometries.

Finally, Theorem 5.3.5 can be also used in a standard way to show a Ferrand-Obata type theorem concerning the conformal quaternionic contact automorphism group, see [98]. Such result was proved in the general context of parabolic geometries admitting regular Cartan connection in [71].

Chapter 6

The quaternionic contact Yamabe problem and the Yamabe constant of the qc spheres

6.1 Introduction

Let M be a compact quaternionic contact manifold M of real dimension $4n + 3$ and homogeneous dimension $Q = 4n + 6$. Thus, we are given an $\mathbb{R}^3$-valued contact form $\eta = \{\eta_1, \eta_2, \eta_3\}$ defining the horizontal distribution H and fundamental two forms $d\eta_{|_H}$ of the quaternionic hermitian structure $(\mathbf{g}, I_1, I_2, I_3)$ on H, $(d\eta_s)_{|_H} = 2\mathbf{g}(I_s., .) = 2\omega_s, s = 1, 2, 3$. A natural question is to determine the qc Yamabe constant of the conformal class $[\eta]$ of η defined as the infimum

$$\lambda(M, [\eta]) = \inf\{\Upsilon(u) : u \in \mathcal{C}^\infty(M), u > 0\}, \qquad (6.1)$$

where $Vol_\eta = \eta_1 \wedge \eta_2 \wedge \eta_3 \wedge (\omega_1)^{2n}$ denotes the volume form determined by η, see (6.9). The qc Yamabe functional of the conformal class of η is defined for $0 < u \in \mathcal{C}^\infty(M)$ by

$$\Upsilon(u) = \left(\int_M \left(4\frac{Q+2}{Q-2}|\nabla u|^2 + S u^2\right) Vol_\eta\right) \bigg/ \left(\int_M u^{2^*} Vol_\eta\right)^{2/2^*}, \qquad (6.2)$$

where Vol_η is the volume form (6.9), ∇ is the Biquard connection of η, S stands for the qc scalar curvature of (M, η) and $|\nabla u|$ is the length of the horizontal gradient, cf. (5.3) and (5.4). This is the so-called *qc Yamabe constant problem*. The question is closely related to the solvability of the qc Yamabe equation (5.25). This is the so called *qc Yamabe problem*, which is the problem of finding all qc structures conformal to a given structure η (of constant qc scalar curvature) which also have constant qc scalar curvature. Taking the conformal factor in the form $\bar{\eta} = u^{4/(Q-2)}\eta$, $Q = 4n + 6$, turns (5.25) into the equation

$$\mathcal{L}u \equiv 4\frac{Q+2}{Q-2}\triangle u - S u = -\overline{S} u^{2^*-1}, \qquad (6.3)$$

157

where $\triangle$ is the horizontal sub-Laplacian, $\triangle u = tr_H^g (\nabla du)$, cf. (5.4), S and $\overline{S}$ are the qc scalar curvatures correspondingly of (M, η) and $(M, \bar{\eta})$, $\bar{\eta} = u^{4/(Q-2)}\eta$, and $2^* = 2Q/(Q-2)$. A natural question is to find all solutions of the qc Yamabe equation (6.3). As usual the two fundamental problems are related by noting that on a compact quaternionic contact manifold M with a fixed conformal class $[\eta]$ the qc Yamabe equation characterizes the non-negative extremals of the qc Yamabe functional.

One of the goals of this chapter is the determination of $\lambda(S^{4n+3}, [\bar{\eta}])$, where $\bar{\eta}$ is the standard qc form (5.30) on the unit sphere S^{4n+3}. In addition, the proof yields *all* qc forms in the conformal class $[\bar{\eta}]$ which achieve $\lambda(S^{4n+3})$. In particular, we shall determine the best constant in the L^2 Folland-Stein inequality (0.4) and characterize *all* functions for which equality holds in the case of the quaternionic Heisenberg group.

The second goal is to solve the qc Yamabe problem for the standard qc structure on the *seven dimensional* sphere and quaternionic Heisenberg groups. We shall see that, as conjectured in [75], all solutions of the Yamabe equations are given by those that realize the Yamabe constant of the sphere or the best constant in the Folland-Stein inequality. In short, we shall prove the next Theorems.

Theorem 6.1.1. [96]

a) *Let* $G(\mathbb{H}) = \mathbb{H}^n \times Im\,\mathbb{H}$ *be the quaternionic Heisenberg group as described in Section 4.3.4. The best constant in the* L^2 *Folland-Stein embedding inequality* (1.1), *where the horizontal gradient is taken using the vector fields* (4.124) *is*

$$S_2 = \frac{\left[2^{2n+6}\sigma_{4n+4}\right]^{-1/(4n+6)}}{\sqrt{2n(n+1)}},$$

Here, $\sigma_{4n+4} = 2\pi^{2n+2}/(2n+1)!$ *is the volume of the unit sphere* $S^{4n+3} \subset \mathbb{R}^{4n+4}$. *The non-negative functions for which* (1.1) *becomes an equality when* $p = 2$ *and* $G = G(\mathbb{H})$ *are given by the functions of the form*

$$F = \gamma\left[(1 + |q|^2)^2 + |\omega|^2\right]^{-(n+1)}, \qquad \gamma = const, \tag{6.4}$$

and all functions obtained from F *by translations* (1.28) *and dilations* (1.29).

b) *The qc Yamabe constant of the standard qc structure of the sphere is*

$$\lambda(S^{4n+3}, [\bar{\eta}]) = 16\,n(n+2)\left[(2n)!\,\sigma_{4n+4}\right]^{1/(2n+3)}. \tag{6.5}$$

The proof, which can be found in Section 6.7, relies on the old idea of center of mass of Szegö [158] and Hersch [89]. We remind that in the Folland-Stein inequality the integration is performed using a Haar measure, while the qc Yamabe constant employs the natural volume form (6.9). The relation between the two integrals on the qc Heisenberg group G ($\mathbb{H}$) is given in (6.64).

Theorem 6.1.2. [95]

a) Let $\tilde{\eta} = \frac{1}{2h}\eta$ be a conformal deformation of the standard qc structure $\tilde{\eta}$ on the quaternionic unit sphere S^7. If η has constant qc-scalar curvature, then up to a multiplicative constant η is obtained from $\tilde{\eta}$ by a conformal quaternionic contact automorphism. In particular, $\lambda(S^7) = 48\,(4\pi)^{1/5}$ and this minimum value is achieved only by $\tilde{\eta}$ and its images under conformal quaternionic contact automorphisms.

b) Let G ($\mathbb{H}$) = $\mathbb{H} \times Im\,\mathbb{H}$ be the seven dimensional quaternionic Heisenberg group. A solution of the Yamabe equation

$$\left(T_1^2 + X_1^2 + Y_1^2 + Z_1^2\right)u = -u^{3/2} \qquad (6.6)$$

is given by the function

$$\bar{u} = 2^{10}\left[(1+|q|^2)^2 + |\omega|^2\right]^{-2}, \qquad (6.7)$$

where the left invariant vector fields are those in (4.124). Any other non-negative entire solution of the Yamabe equation is obtained from $\bar{u}$ by translations (6.50) and dilations (6.51).

We note that $2^* - 1 = (Q+2)/(Q-2) = 3/2$ when $n = 1$.

The proof of Theorem 6.1.2 is given in Section 6.6 and relies on the divergence formula established in Theorems 6.5.1 and 6.2.5.

6.2 Some background

In this section we present key results from [94] necessary for the proof of the main theorems.

6.2.1 The qc normal frame

The next Lemma established in [94] is an application of a standard result in differential geometry is very useful and simplifies the calculations.

Lemma 6.2.1. *In a neighborhood of any point $p \in M^{4n+3}$ and a qc-orthonormal basis*

$$\{X_1(p), X_2(p) = I_1 X_1(p) \ldots, X_{4n}(p) = I_3 X_{4n-3}(p), \xi_1(p), \xi_2(p), \xi_3(p)\}$$

of the tangential space at p there exists a qc-orthonormal frame field

$$\{X_1, X_2 = I_1 X_1, \ldots, X_{4n} = I_3 X_{4n-3}, \xi_1, \xi_2, \xi_3\}$$

such that $X_{a|p} = X_a(p), \xi_{s|p} = \xi_s(p), a = 1, \ldots, 4n$ and the connection 1-forms of the Biquard connection are all zero at the point p:

$$(\nabla_{X_a} X_b)_{|p} = (\nabla_{\xi_i} X_b)_{|p} = (\nabla_{X_a} \xi_t)_{|p} = (\nabla_{\xi_t} \xi_s)_{|p} = 0, \qquad (6.8)$$

for $a, b = 1, \ldots, 4n, s, t, r = 1, 2, 3$.

In particular,

$$((\nabla_{X_a} I_s) X_b)_{|p} = ((\nabla_{X_a} I_s) \xi_t)_{|p} = ((\nabla_{\xi_t} I_s) X_b)_{|p} = ((\nabla_{\xi_t} I_s) \xi_r)_{|p} = 0.$$

Proof. Since ∇ preserves the splitting $H \oplus V$ we can apply the standard arguments for the existence of a normal frame with respect to a metric connection (see e.g. [172]). We sketch the proof for completeness.

Let $\{\tilde{X}_1, \ldots, \tilde{X}_{4n}, \tilde{\xi}_1, \tilde{\xi}_2, \tilde{\xi}_3\}$ be a qc-orthonormal basis around p such that $\tilde{X}_{a|p} = X_a(p)$, $\tilde{\xi}_{i|p} = \xi_i(p)$. We want to find a modified frame $X_a = o_a^b \tilde{X}_b$, $\xi_i = o_i^j \tilde{\xi}_j$, which satisfies the normality conditions of the lemma.

Let ϖ be the $sp(n) \oplus sp(1)$-valued connection 1-forms with respect to $\{\tilde{X}_1, \ldots, \tilde{X}_{4n}, \tilde{\xi}_1, \tilde{\xi}_2, \tilde{\xi}_3\}$,

$$\nabla \tilde{X}_b = \varpi_b^c \tilde{X}_c, \quad \nabla \tilde{\xi}_s = \varpi_s^t \tilde{\xi}_t.$$

Let $\{x^1, \ldots, x^{4n+3}\}$ be a coordinate system around p such that

$$\frac{\partial}{\partial x^a}(p) = X_a(p), \quad \frac{\partial}{\partial x^{4n+t}}(p) = \xi_t(p), \quad a = 1, \ldots, 4n, \quad t = 1, 2, 3.$$

One can easily check that the matrices

$$o_a^b = exp\left(-\sum_{c=1}^{4n+3} \varpi_a^b \left(\frac{\partial}{\partial x^c}\right)_{|p} x^c \right) \in Sp(n),$$

$$o_t^s = exp\left(-\sum_{c=1}^{4n+3} \varpi_t^s \left(\frac{\partial}{\partial x^c}\right)_{|p} x^c \right) \in Sp(1)$$

are the desired matrices making the identities (6.8) true.

Now, the last identity in the lemma is a consequence of the fact that the choice of the orthonormal basis of V does not depend on the action of $SO(3)$ on V combined with (4.10). $\qquad \square$

Definition 6.2.2. We refer to the orthonormal frame constructed in Lemma 6.2.1 as a *qc-normal frame.*

6.2.2 *Horizontal divergence theorem*

Let (M, η) be a quaternionic contact manifold with a fixed globally defined contact form η. For a fixed $s \in \{1, 2, 3\}$ the form

$$Vol_\eta = \eta_1 \wedge \eta_2 \wedge \eta_3 \wedge \omega_s^{2n} \qquad (6.9)$$

is a volume form and is independent of s.

We define the (horizontal) divergence of a horizontal vector field/one-form $\sigma \in \Lambda^1(H)$ by

$$\nabla^* \sigma = tr|_H \nabla \sigma = \sum_{a=1}^{4n} (\nabla_{e_\alpha} \sigma)(e_\alpha). \qquad (6.10)$$

Clearly the horizontal divergence does not depend on the basis and is an $Sp(n)Sp(1)$-invariant.

For any horizontal 1-form $\sigma \in \Lambda^1(H)$ we denote with $\sigma^\#$ the corresponding horizontal vector field via the horizontal metric defined with the equality $\sigma(X) = g(\sigma^\#, X)$.

It is justified to call the function $\nabla^* \sigma$ divergence of σ in view of the following Proposition.

Proposition 6.2.3. *Let (M, η) be a quaternionic contact manifold of dimension $(4n+3)$ and $\eta \wedge \omega_s^{2n-1} \overset{def}{=} \eta_1 \wedge \eta_2 \wedge \eta_3 \wedge \omega_s^{2n-1}$.*

For any horizontal 1-form $\sigma \in \Lambda^1(H)$ we have

$$d(\sigma^\# \lrcorner (Vol_\eta)) = -(\nabla^* \sigma) Vol_\eta .$$

Therefore, if M is compact,

$$\int_M (\nabla^* \sigma) Vol_\eta = 0.$$

Proof. We work in a qc-normal frame at a point $p \in M$ constructed in Lemma 6.2.1. Let e^a be the horizontal 1-form corresponding to the vector field e_a, $e^a(X) = g(X, e_a)$. Since σ is horizontal, we have $\sigma^\# = g(\sigma^\#, e_a)e_a$. Therefore, we calculate

$$\sigma^\# \lrcorner (Vol_\eta) = \sum_{a=1}^{4n} (-1)^a g(\sigma^\#, e_a) \eta \wedge e^{a_1} \wedge \cdots \wedge \hat{e^a} \wedge \cdots \wedge e^{4n},$$

where $\hat{e^a}$ means that the 1-form e^a is missing in the above wedge product.

The exterior derivative of the above expression gives

$$d(\sigma^{\#} \lrcorner (Vol_\eta)) = -\sum_{a=1}^{4n} e_a g(\sigma^{\#}, e_a) Vol_\eta$$

$$+ \sum_{a=1}^{4n} \sum_{s=1}^{3} g(\sigma^{\#}, e_a) d\eta_s(\xi_s, e_a) Vol_\eta \quad \text{mod} \quad \text{sign}$$

$$+ \sum_{a=1, b=1}^{4n} g(\sigma^{\#}, e_a) de^b(e_b, e_a) Vol_\eta \quad \text{mod} \quad \text{sign}$$

$$= -(\nabla^* \sigma) Vol_\eta. \quad (6.11)$$

Indeed, since the Biquard connection preserves the metric, the right hand side of the first line can be handled as follows $e_a g(\sigma^{\#}, e_a) = g(\nabla_{e_a} \sigma^{\#}, e_a) + g(\sigma^{\#}, \nabla_{a_a} e_a)$ which evaluated at the point p gives

$$e_a g(\sigma^{\#}, e_a)_{|_p} = g(\nabla_{e_a} \sigma^{\#}, e_a)_{|_p}) = ((\nabla_{e_a} \sigma) e_a)_{|_p}.$$

The second line of (6.11) is equal to zero because of the definition of the Reeb vector fields, (4.7). The third line vanishes because of the following sequence of identities

$$de^b(e_b, e_a)_{|_p} = e^b([e_b, e_a])_{|_p} = e^b(\nabla_{e_b} e_a - \nabla_{e_a} e_b - T(e_b, e_a))_{|_p} = 0$$

since $T(e_b, a_a)$ is a vertical vector field. This proves the first formula. When the manifold is compact, an application of the Stoke's theorem completes the proof. $\quad\quad\quad\square$

We note that the integral formula of the above theorem was essentially proved in [[168], Proposition 2.1].

Remark 6.2.4. From Remark 5.2.2 it follows that the volume form $Vol_{\tilde{\eta}}$ of the standard qc form (5.30) on the unit sphere is

$$Vol_{\tilde{\eta}} = \kappa_{\tilde{\eta}} d\sigma, \quad\quad \kappa_{\tilde{\eta}} = 2^{2n+3}(2n)! \quad (6.12)$$

where $d\sigma$ is the Riemannian volume form of the round unit sphere in $\mathbb{H}^{n+1}$.

6.2.3   *Conformal transformations of the quaternionic Heisenberg group preserving the vanishing of the torsion*

Studying conformal deformations of qc structures preserving the qc-Einstein condition, we describe explicitly all global functions on the quaternionic Heisenberg group sending conformally the standard flat qc structure to another qc-Einstein structure in [94].

Theorem 6.2.5. *[94] Let* $\Theta = \frac{1}{2h}\tilde{\Theta}$ *be a conformal deformation of the standard qc-structure* $\tilde{\Theta}$ *on the quaternionic Heisenberg group* $G(\mathbb{H})$. *If* Θ *is also qc-Einstein, then up to a left translation the function* h *is given by*

$$h = c\left[\left(1 + \nu|q|^2\right)^2 + \nu^2(x^2 + y^2 + z^2)\right],$$

where c *and* ν *are positive constants. All functions* h *of this form have this property.*

Here we sketch the proof formulating only the crucial steps in a sequence of lemmas and propositions, and refer to [94] for a complete proof.

We start with a Proposition in which we determine the vertical Hessian of h.

Proposition 6.2.6. *If* h *satisfies* (5.26) *on* $G(\mathbb{H})$ *then we have the relations*

$$\xi_1^2(h) = \xi_2^2(h) = \xi_3^2(h) = 8\mu_o, \quad \xi_s\xi_t(h) = 0, \quad s \neq t = 1, 2, 3, \qquad (6.13)$$

where $\mu_o > 0$ *is a constant. In particular,*

$$\begin{aligned} h(q, \omega) &= g(q) \\ &+ \mu_o\left[(x + x_o(q))^2 + (y + y_o(q))^2 + (z + z_o(q))^2\right] \quad (6.14) \end{aligned}$$

for some real valued functions g, x_o y_o *and* z_o *on* $\mathbb{H}^n$. *Furthermore we have*

$$T_a Z_a X_a^2(h) = T_a Z_a Y_a^2(h) = 0, \quad T_a^2 \xi_j(h) = 0.$$

In view of Proposition 6.2.6, we define $h = g + \mu_o f$, where

$$f = (x + x_o(q))^2 + (y + y_o(q))^2 + (z + z_o(q))^2. \qquad (6.15)$$

The following simple Lemma is one of the keys to integrating the system (5.26) and (5.27).

Lemma 6.2.7. *Let* X *and* Y *be two parallel horizontal vectors*

a) If $\omega_s(X, Y) = 0$ *then*

$$4XYh - 2h^{-1}\left[dh(X)\,dh(Y) + \sum_{s=1}^{3} dh(I_s X)\,dh(I_s Y)\right] = \lambda g(X, Y).$$

$$(6.16)$$

b) If $g(X, Y) = 0$ *then*

$$2\,XYh \;-\; h^{-1}\{dh(X)\,dh(Y) \;+\; \sum_{s=1}^{3} dh(I_sX)\,dh(I_sY)\}$$

$$= 2\sum_{s=1}^{3}\{\,(\xi_s h)\,\omega_s(X, Y)\,\}. \quad (6.17)$$

c) If $g(X, Y) = \omega_s(X, Y) = 0$ *for* $s = 1, 2, 3$ *we have for any* $t \in \{1, 2, 3\}$ *that*

$$XY\,(\xi_t h) = 0,$$

$$8\,XYh \;=\; \mu_o\{(X\xi_t f)\,(Y\xi_t f) \;+\; \sum_{s=1}^{3}(I_sX\,\xi_t f)\,(I_sY\,\xi_t f)\,\}. \quad (6.18)$$

In order to see that after a suitable translation the functions x_o y_o and z_o can be made equal to zero we need the following proposition.

Proposition 6.2.8. *If* h *satisfies* (5.26) *and* (5.27) *on* $G\,(\mathbb{H})$ *then we have*

a) For $s \in \{1, 2, 3\}$ *and* i, j, k *a cyclic permutation of* $1, 2, 3$

$$T_a\,T_b\,(\xi_s h) = (I_iT_a)\,(I_iT_b)\,(\xi_s h) = 0 \quad \forall\, a,\, b$$

$$(I_iT_a)\,T_b\,(\xi_s h) = (I_iT_a)\,(I_iT_b)\,(\xi_s h) = 0,\; a \neq b$$

$$(I_j\,T_a)\,T_a\,(\xi_s\,h) = -\,T_a\,(I_j\,T_a)\,(\xi_s h) = 8\,\delta_{sj}\,\mu_o \qquad (6.19)$$

$$(I_j\,T_a)\,(I_i\,T_a)\,(\xi_s h) = -\,(I_i\,T_a)\,(I_j\,T_a)\,(\xi_s h) = 8\,\delta_{sk}\,\mu_o\,,$$

i.e., the horizontal Hessian of a vertical derivative of h *is determined completely.*

b) There is a point $(q_0, \omega_o) \in G\,(\mathbb{H})$, $q_o = (q_o^1, q_o^2, \ldots, q_o^n) \in \mathbb{H}^n$ *and* $\omega = ix_o + jy_o + kz_o \in Im(\mathbb{H})$, *such that,*

$$ix_o(q) \;+\; jy_o(q) \;+\; kz_o(q) = w_o \;+\; 2\,Im\,q_o\,\bar{q}.$$

So far we have proved that if h satisfies the system (5.26) and (5.27) on $G\,(\mathbb{H})$ then, in view of the translation invariance of the system, after a suitable translation we have

$$h(q, \omega) = g(q) \;+\; \mu_o\,(x^2 \;+\; y^2 \;+\; z^2).$$

The goal is to show that $g(q) = (b + 1 + \sqrt{\mu}_o\,|q|^2)^2$. The next result completes the proof of Theorem 6.2.5.

Proposition 6.2.9. *If* h *satisfies the system* (5.26) *and* (5.27) *on* $G\,(\mathbb{H})$ *then after a suitable translation we have*

$$g(q) = (b + 1 + \sqrt{\mu}_o\,|q|^2)^2, \qquad b + 1 > 0.$$

6.3 Constant qc scalar curvature and the divergence formula

In this section we investigate in details qc conformal deformations of positive 3-Sasakian manifolds which lead to a qc manifolds with constant qc scalar curvature.

We consider the following horizontal 1-forms A_s, $A = A_1 + A_2 + A_3$ defined by

$$A_i(X) = \omega_i([\xi_j, \xi_k], X), \qquad (6.20)$$

which, in view of equation (4.74) can be written in the form

$$A_i(X) = -\rho_i(I_j X, \xi_k) = \rho_i(I_k X, \xi_j) = -T(\xi_j, \xi_k, I_i X). \qquad (6.21)$$

An immediate consequence of (4.108) and (4.50) is the following

Corollary 6.3.1. *On a $4n+3$-dimensional qc manifold of constant qc scalar curvature the next formulas hold*

$$A_i(X) = \frac{2(n+1)}{3(2n+1)} \nabla^* U(X) + \frac{4n+1}{12(2n+1)} \nabla^* T^0(X)$$
$$+ \frac{1}{4(2n+1)} (\nabla_{e_a} T^0)(I_i e_a, I_i X). \quad (6.22)$$

$$A(X) = \frac{2(n+1)}{2n+1} \nabla^* U(X) + \frac{n}{2n+1} \nabla^* T^0(X). \qquad (6.23)$$

We obtain from the contracted Bianchi identity (4.121) and (6.23) the next result established in [[94], Theorem 4.8].

Theorem 6.3.2. *[94] On a $(4n+3)$-dimensional qc manifold with constant qc-scalar curvature we have the formulas*

$$\nabla^* T^0 = (n+2)A, \qquad \nabla^* U = \frac{1-n}{2} A. \qquad (6.24)$$

Hereafter in this section (M, g, η) will be a qc manifold with globally defined qc structure η and constant positive qc-scalar curvature $Scal$. We may assume $Scal = 16n(n+2)$ since we can always use a qc homothety to arrange this constant.

Lemma 6.3.3. *Let h be a positive smooth function on a qc manifold (M, g, η) with a constant positive qc-scalar curvature $Scal = 16n(n+2)$*

and $\bar{\eta} = \frac{1}{2h}\eta$ a conformal deformation of the qc structure η. If $\bar{\eta}$ is a positive 3-Sasakian structure, then we have the formulas

$$A_1(X) = -\frac{1}{2}h^{-2}dh(X) - \frac{1}{2}h^{-3}|\nabla h|^2 dh(X)$$

$$- \frac{1}{2}h^{-1}\Big(\nabla dh(I_2 X, \xi_2) + \nabla dh(I_3 X, \xi_3)\Big)$$

$$+ \frac{1}{2}h^{-2}\Big(dh(\xi_2)\,dh(I_2 X) + dh(\xi_3)\,dh(I_3 X)\Big)$$

$$+ \frac{1}{4}h^{-2}\Big(\nabla dh(I_2 X, I_2 \nabla h) + \nabla dh(I_3 X, I_3 \nabla h)\Big). \quad (6.25)$$

The expressions for A_2 and A_3 can be obtained from the above formula by a cyclic permutation of $(1,2,3)$. Thus, we have also

$$A(X) = -\frac{3}{2}h^{-2}dh(X) - \frac{3}{2}h^{-3}|\nabla h|^2 dh(X) - h^{-1}\sum_{s=1}^{3} \nabla dh(I_s X, \xi_s)$$

$$+ h^{-2}\sum_{s=1}^{3} dh(\xi_s)\,dh(I_s X) + \frac{1}{2}h^{-2}\sum_{s=1}^{3} \nabla dh(I_s X, I_s \nabla h).$$

Proof. First we calculate the $sp\,(1)$-connection 1-forms of the Biquard connection ∇. For a positive 3-Sasakian structure we derive from (4.134), applying (4.92) and (4.93), that

$$d\bar{\eta}_i(\bar{\xi}_j, \bar{\xi}_k) = 2, \quad \bar{\xi}_i \lrcorner d\bar{\eta}_i = 0, \quad \bar{\alpha}_s(\bar{\xi}_t) = -2\delta_{st}$$

and the qc-scalar curvature $\overline{Scal} = 16n(n+2)$.

Then (5.2), (4.92), and (4.93) yield

$$2d\eta_i(\xi_j, \xi_k) = 2h^{-1} + h^{-2}\|dh\|^2,$$

$$\alpha_i(X) = -h^{-1}dh(I_i X), \quad \alpha_i(\xi_j) = -h^{-1}dh(\xi_k) = -\alpha_j(\xi_i), \quad (6.26)$$

$$4\alpha_i(\xi_i) = -4 - 2h^{-1} - h^{-2}\|dh\|^2.$$

From the 3-Sasakian assumption the commutators are $[\bar{\xi}_i, \bar{\xi}_j] = -2\bar{\xi}_k$. Thus, for $X \in H$ taking also into account (5.2) we have

$$g([\bar{\xi}_1, \bar{\xi}_2], I_3 X) = -2g(\bar{\xi}_3, I_3 X) = -2g(2h\xi_3 + I_3\nabla h, I_3 X)$$

$$= -2dh(X).$$

Therefore, using again (5.2), we obtain

$$-2dh(X) = g([\bar{\xi}_1, \bar{\xi}_2], I_3 X)$$

$$= g\big([2h\xi_1 + I_1\nabla h, 2h\xi_2 + I_2\nabla h], I_3 X\big)$$

$$= -4h^2 A_3(X) + 2hg([\xi_1, I_2\nabla h], I_3 X)$$

$$+ 2hg([I_1\nabla h, \xi_2], I_3 X) + g([I_1\nabla h, I_2\nabla h], I_3 X). \quad (6.27)$$

The last three terms we evaluate as follows. The first equals

$$g\left([\xi_1, I_2\nabla h], I_3 X\right) = g\left((\nabla_{\xi_1} I_2)\nabla h + I_2\nabla_{\xi_1}\nabla h, I_3 X\right)$$
$$- g\left(T(\xi_1, I_2\nabla h), I_3 X\right) = -\alpha_3(\xi_1)\, dh(I_2 X) + \alpha_1(\xi_1)\, dh(X)$$
$$- \nabla dh\left(\xi_1, I_1 X\right) - g\left(T(\xi_1, I_2\nabla h), I_3 X\right),$$

where we use (4.10) and the fact that ∇ preserves the splitting $H \oplus V$. We obtain for the second term that

$$g\left([I_1\nabla h, \xi_2], I_3 X\right) = \alpha_2(\xi_2)\, dh(X) + \alpha_3(\xi_2)\, dh(I_1 X)$$
$$- \nabla dh\left(\xi_2, I_2 X\right) - g\left(T(I_1\nabla h, \xi_2), I_3 X\right).$$

Working similarly, we obtain for the third term the equalities

$$g\left([I_1\nabla h, I_2\nabla h], I_3 X\right) = -\alpha_3(I_1\nabla h)\, dh(I_2 X) + \alpha_1(I_1\nabla h)\, dh(X)$$
$$- \nabla dh\left(I_1\nabla h, I_1 X\right) + \alpha_2(I_2\nabla h)\, dh(X)$$
$$+ \alpha_3(I_2\nabla h)\, dh(I_1 X) - \nabla dh\left(I_2\nabla h, I_2 X\right).$$

Next we apply (6.26) to the last three equalities, then substitute their sum into (6.27), after which we use the commutation relations (5.11) to obtain the following identity

$$4h^2 A_3(X) = (-4h + h^{-1}\|\nabla h\|^2)\, dh(X)$$
$$- 2h\left[\nabla dh\left(I_1 X, \xi_1\right) + \nabla dh\left(I_2 X, \xi_2\right)\right]$$
$$- \left[\nabla dh\left(I_1 X, I_1\nabla h\right)) + \nabla dh\left(I_2 X, I_2\nabla h\right)\right]$$
$$+ 2\left[dh(\xi_1)\, dh(I_1 X) + dh(\xi_2)\, dh(I_2 X) + 2\, dh(\xi_3)\, dh(I_3 X)\right]$$
$$+ 2h\left[T(\xi_1, I_1 X, \nabla h) + T(\xi_2, I_2 X, \nabla h)\right]$$
$$- 2h\left[T(\xi_1, I_2 X, I_3\nabla h) - T(\xi_2, I_1 X, I_3\nabla h)\right]. \quad (6.28)$$

Using (4.20), (4.51) and (4.53), we obtain that the torsion endomorphism of the Biquard connection is given by

$$T(\xi_s, X, Y) = -\frac{1}{4}\left(T^0(I_s X, Y) + T^0(X, I_s Y)\right) + U(I_s X, Y). \quad (6.29)$$

The equation (6.29) together with (4.50) and (4.52) help us to represent the last two lines in (6.28) in terms of the tensors T^0 and U as follows

$$T(\xi_1, I_1 X, \nabla h) + T(\xi_2, I_2 X, \nabla h) - T(\xi_1, I_2 X, I_3\nabla h) + T(\xi_2, I_1 X, I_3\nabla h)$$
$$= T^0(X, \nabla h) + T^0(I_3 X, I_3\nabla h) - 8U(X, \nabla h),$$

which allows us to rewrite (6.28) in the form

$$
\begin{aligned}
4A_3(X) \;=\; & (-4h^{-1} + h^{-3}\|\nabla h\|^2)\,dh(X) \\
& - 2\,h^{-1}\big[\nabla dh\,(I_1 X, \xi_1) + \nabla dh\,(I_2 X, \xi_2)\big] \\
& + 2\,h^{-2}\big[dh(\xi_1)\,dh(I_1 X) + dh(\xi_2)\,dh(I_2 X) + 2\,dh(\xi_3)\,dh(I_3 X)\big] \\
& \quad h^{-2}\big[\nabla dh\,(I_1 X, I_1\nabla h) + \nabla dh\,(I_2 X, I_2\nabla h)\big] \\
& + h^{-1}\big[2T^0(X,\nabla h) + 2T^0(I_3 X, I_3\nabla h) - 8U(X,\nabla h)\big]. \quad (6.30)
\end{aligned}
$$

Using (5.12) the T^0 part of the torsion in (6.28) can be expressed by h as follows,

$$
\begin{aligned}
2\,T^0(\nabla h, X) \;&+\; 2\,T^0(I_3\nabla h, I_3 X) \\
&= -2h^{-1}\Big\{[\nabla dh]_{[-1]}(\nabla h, X) + [\nabla dh]_{[-1]}(I_3\nabla h, I_3 X)\Big\} \\
&\quad - 2h^{-1}\sum_{s=1}^{3}\Big\{dh(\xi_s)\big[g(I_s\nabla h, X) + g(I_s I_3\nabla h, I_3 X)\big]\Big\} \\
&= -h^{-1}\Big\{\nabla dh\,(\nabla h, X) + \nabla dh\,(I_3\nabla h, I_3 X)\Big\} \\
&\quad + h^{-1}\Big\{\nabla dh\,(I_1\nabla h, I_1 X) + \nabla dh\,(I_2\nabla h, I_2 X)\Big\} \\
&\qquad\qquad\qquad\qquad\qquad + 4h^{-1}\,dh(\xi_3)\,dh(I_3 X).
\end{aligned}
$$

Invoking equation (5.11) we can put ∇h in second place in the Hessian terms, thus, proving the formula

$$
\begin{aligned}
2\,T^0(\nabla h, X) \;+\; 2\,T^0(I_3\nabla h, I_3 X) \;=\; & -4h^{-1}\,dh(\xi_3)\,dh(I_3 X) \\
& - h^{-1}\Big\{\nabla dh\,(X,\nabla h) + \nabla dh\,(I_3 X, I_3\nabla h)\Big\} \\
& + h^{-1}\Big\{\nabla dh\,(I_1 X, I_1\nabla h) + \nabla dh\,(I_2 X, I_2\nabla h)\Big\}. \quad (6.31)
\end{aligned}
$$

Similarly, (5.23) and the Yamabe equation (5.25) give

$$8U(\nabla h, X) = -h^{-1}\left\{\nabla dh\,(\nabla h, X) + \sum_{s=1}^{3} \nabla dh\,(I_s\nabla h, I_s X)\right.$$

$$\left. - 2h^{-1}\|\nabla h\|^2 dh(X) - \frac{\Delta h}{n} dh(X) + 2h^{-1}\frac{\|\nabla h\|^2}{n} dh(X)\right\}$$

$$= -h^{-1}\left\{\nabla dh\,(\nabla h, X) + \sum_{s=1}^{3} \nabla dh\,(I_s\nabla h, I_s X)\right\}$$

$$- h^{-1}\left\{-2h^{-1}\|\nabla h\|^2 dh(X) - \frac{2n - 4nh + (n+2)h^{-1}\|\nabla h\|^2}{n} dh(X)\right.$$

$$\left. + 2h^{-1}\frac{\|\nabla h\|^2}{n} dh(X)\right\}$$

$$= -h^{-1}\left\{\nabla dh\,(X, \nabla h) + \sum_{s=1}^{3} \nabla dh\,(I_s X, I_s\nabla h)\right\}$$

$$- h^{-1}\left(-3h^{-1}\|\nabla h\|^2 - 2 + 4h\right)dh(X). \qquad (6.32)$$

Substituting the last two formulas (6.31) and (6.32) into (6.30) gives A_3 in the form of (6.25) written for A_1, cf. the paragraph after (6.25). $\qquad\square$

6.4   Divergence formulas

We shall need the divergences of various vector/1-forms through the almost complex structures, so we start with a general formula valid for any horizontal vector/1-form σ. Let $\{e_1, \ldots, e_{4n}\}$ be an orthonormal basis of H. The divergence of $I_s\sigma$ is

$$\nabla^*(I_s\sigma) \equiv (\nabla_{e_a}(I_s\sigma))(e_a) = -(\nabla_{e_a}\sigma)(I_s e_a) - \sigma((\nabla_{e_a}I_s)e_a),$$

recalling $I_s\sigma(X) = -\sigma(I_s X)$.

We recall that an orthonormal frame

$$\{e_1, e_2 = I_1 e_1, e_3 = I_2 e_1, e_4 = I_3 e_1, \ldots, e_{4n} = I_3 e_{4n-3}, \xi_1, \xi_2, \xi_3\}$$

is a qc-normal frame (at a point) if the connection 1-forms of the Biquard connection vanish (at that point). Lemma 6.2.1 asserts that a qc-normal frame exists at each point of a qc manifold. With respect to a qc-normal frame the above divergence reduces to

$$\nabla^*(I_s\sigma) = -(\nabla_{e_a}\sigma)(I_s e_a).$$

Lemma 6.4.1. *Suppose $(M, g, \mathbb{Q})$ is a quaternionic contact manifold with constant qc-scalar curvature. For any function h we have the following* *formulas*

$$\nabla^* \left(\sum_{s=1}^{3} dh(\xi_s) I_s A_s \right) = \sum_{s=1}^{3} \nabla dh \, (I_s e_a, \xi_s) A_s(e_a)$$

$$\nabla^* \left(\sum_{s=1}^{3} dh(\xi_s) I_s A \right) = \sum_{s=1}^{3} \nabla dh \, (I_s e_a, \xi_s) A(e_a).$$

Proof. Using the identification of the 3-dimensional vector spaces spanned by $\{\xi_1, \xi_2, \xi_3\}$ and $\{I_1, I_2, I_3\}$ with $\mathbb{R}^3$, the restriction of the action of $Sp(n)Sp(1)$ to this spaces can be identified with the action of the group $SO(3)$, i.e., $\xi_s = \sum_{t=1}^{3} \Psi_{st} \bar{\xi}_t$ and $I_s = \sum_{t=1}^{3} \Psi_{st} \bar{I}_t$, with $\Psi \in SO(3)$. One verifies easily that the vectors

$$A, \quad \sum_{s=1}^{3} dh(\xi_s) I_s A_s = -\sum_{i=1}^{3} dh(\xi_i)[\xi_j, \xi_k] \quad \text{and} \quad \sum_{s=1}^{3} dh(\xi_s) I_s A$$

are $Sp(n)Sp(1)$ invariant on $\mathbb{H}$, for example $\bar{A} = (det\Psi) A = A$. Thus, it is sufficient to compute their divergences in a qc-normal frame. To avoid the introduction of new variables, in this proof, we shall assume that $\{e_1, \ldots, e_{4n}, \xi_1, \xi_2, \xi_3\}$ is a qc-normal frame.

To prove the first formula we show $\nabla^*(I_s A_s) = 0$.

The formula (4.74) yields

$$-\nabla^*(I_3 A_3) = \nabla^*[\xi_1, \xi_2] = (\nabla_{e_a} \rho_3)(I_1 e_a, \xi_1).$$

On the other hand, taking the trace into (4.106) and using that the torsion endomorphism of the Biquard connection is traceless, we calculate

$$0 = R(\xi_1, \xi_2, e_b, e_b) = -(\nabla_{e_b} \rho_3)(I_1 e_b, \xi_1)$$
$$- T(\xi_j, e_b, e_a) T(\xi_i, e_a, e_b) + T(\xi_j, e_a, e_b) T(\xi_i, e_b, e_a)$$
$$= -(\nabla_{e_b} \rho_3)(I_1 e_b, \xi_1).$$

Hence $\nabla^*(I_3 A_3) = 0$. The equalities $\nabla^*(I_1 A_1) = \nabla^*(I_2 A_2) = 0$ with respect to a qc-normal frame can be obtained similarly. Hence, the first formula in Lemma 6.4.1 follows.

We are left with proving the second divergence formula. For that purpose it is sufficient to show $\nabla^*(I_s A) = 0$.

From the definition of A, we have

$$I_2 A = -[\xi_3, \xi_1] + I_3[\xi_3, \xi_2] + I_1[\xi_1, \xi_2]. \tag{6.33}$$

Since the qc scalar curvature is constant, we calculate taking suitable traces from (4.106) and applying (4.76) and (4.74) that

$$0 = 4n(\rho_1(\xi_1, \xi_2) + \rho_3(\xi_3 \xi_2))$$
$$= R(\xi_1, \xi_2, e_b, I_1 e_b) - R(\xi_2, \xi_3, e_b, I_3 e_b)$$
$$= (\nabla_{e_b} \rho_3)(e_b, \xi_1) - T(\xi_2, e_b, e_a)T(\xi_1, e_a, I_1 e_b) + T(\xi_2, e_a, I_1 e_b)T(\xi_1, e_b, e_a)$$
$$+ (\nabla_{e_b} \rho_1)(I_1 e_b, \xi_2) + T(\xi_3, e_b, e_a)T(\xi_2, e_a, I_3 e_b) - T(\xi_3, e_a, I_3 e_b)T(\xi_2, e_b, e_a)$$
$$= \nabla^*(I_1[\xi_1, \xi_2] + I_3[\xi_3, \xi_2]) - T(\xi_2, e_b, e_a)\Big(T(\xi_1, e_a, I_1 e_b) + T(\xi_1, I_1 e_a, e_b)\Big)$$
$$- T(\xi_2, e_b, e_a)\Big(T(\xi_3, e_a, I_3 e_b) + T(\xi_3, I_3 e_a, e_b)\Big)$$
$$= \nabla^*(I_1[\xi_1, \xi_2] + I_3[\xi_3, \xi_2]) \quad (6.34)$$

because the terms involving the torsion in (6.34) vanish. Indeed, taking into account that the scalar product af symmetric and skew-symmetric tensors vanishes, using (4.20), (4.43), (4.50), (4.51) and (4.53) we calculate

$$T(\xi_2, e_b, e_a)\Big(T(\xi_1, e_a, I_1 e_b) + T(\xi_1, I_1 e_a, e_b)\Big)$$
$$+ T(\xi_2, e_b, e_a)\Big(T(\xi_3, e_a, I_3 e_b) + T(\xi_3, I_3 e_a, e_b)\Big)$$
$$= 2T^0(\xi_2, e_b, e_a)\Big(T^0(\xi_1, I_1 e_a, e_b) + T^0(\xi_3, I_3 e_a, e_b)\Big)$$
$$= -\frac{1}{2}\Big(T^0(I_2 e_b, e_a) + T^0(e_b, I_2 e_a))\Big)\frac{1}{4}\Big[3T^0(e_a, e_b) + T^0(I_2 e_a, I_2 e_b)\Big]$$
$$= -\frac{1}{4}\Big(T^0(I_2 e_b, e_a) + T^0(e_b, I_2 e_a)\Big)T^0(e_a, e_b) = 0$$

using suitable switch to the basis $\{I_2 e_a : a = 1, \ldots, 4n\}$ to obtain the first equality in the fifth line. To get the last equality we rely on the identity

$$T^0(I_2 e_b, e_a)T^0(e_b, e_a) = 0$$

following, for example, from the equality

$$T^0(I_2 e_b, e_a)T^0(e_b, e_a) = -T^0(e_b, e_a)T^0(I_2 e_b, e_a).$$

Combining (6.34) with $\nabla^*(I_2 A_2) = 0$ we conclude from (6.33) that $\nabla^*(I_2 A) = 0$. Similarly, we derive $\nabla^*(I_1 A) = \nabla^*(I_3 A) = 0$ which proofs the second formula in Lemma 6.4.1. $\qquad\square$

We shall also need the following one-forms

$$D_s(X) = -\frac{1}{2}h^{-1}\Big[T^0(X, \nabla h) + T^0(I_s X, I_s \nabla h)\Big]. \quad (6.35)$$

For simplicity, using the musical isomorphism, we will denote with D_s the corresponding (horizontal) vector fields, for example $g(D_s, X) = D_s(X)$.

We also set

$$D = D_1 + D_2 + D_3 = -h^{-1}T^0(X, \nabla h), \qquad (6.36)$$

where we used (4.50) to obtain the second equality in (6.36).

Lemma 6.4.2. *Suppose* (M, η) *is a quaternionic contact manifold with constant qc-scalar curvature* $Scal = 16n(n+2)$. *Suppose* $\bar{\eta} = \frac{1}{2h}\eta$ *has vanishing* $[-1]$*-torsion component* $\overline{T}^0 = 0$.

a) We have

$$D(X) = \frac{1}{4}h^{-2}\left(3\,\nabla dh(X, \nabla h) - \sum_{s=1}^{3}\nabla dh(I_s X, I_s \nabla h)\right)$$

$$+ h^{-2}\sum_{s=1}^{3}dh(\xi_s)\,dh(I_s X).$$

b) The divergence of D satisfies

$$\nabla^* D = |T^0|^2 - h^{-1}g(dh, D) - h^{-1}(n+2)\,g(dh, A).$$

Proof. a) The formula for D follows immediately from (5.12).

b) We work in a qc-normal frame. Since the qc scalar curvature is assumed to be constant we use (6.24) to find

$$\nabla^* D = -h^{-1}\,dh(e_a)D(e_a)$$
$$- h^{-1}\nabla^* T^0(\nabla h) - h^{-1}T^0(e_a, e_b)\,\nabla dh(e_a, e_b)$$
$$= -h^{-1}\,dh(e_a)D(e_a) - h^{-1}(n+2)\,dh(e_a)A(e_a) - g(T^0, h^{-1}\,\nabla dh)$$
$$= |T^0|^2 - h^{-1}dh(e_a)D(e_a) - h^{-1}(n+2)\,dh(e_a)A(e_a),$$

applying (5.12) in the last equality. $\qquad\square$

Let us also consider the following horizontal one-forms (and corresponding horizontal vector fields)

$$F_s(X) = -h^{-1}T^0(X, I_s\nabla h). \qquad (6.37)$$

From the definition of D_s, (6.35), we find

$$- D_i(I_i X) + D_j(I_i X) + D_k(I_i X)$$
$$= -\frac{h^{-1}}{2}\left[T^0(X, I_i\nabla h) + T^0(I_i X, \nabla h) - T^0(I_k X, I_j\nabla h) + T^0(I_j X, I_k\nabla h)\right]$$
$$= -h^{-1}T^0(X, I_i\nabla h) = F_i(X),$$

where we applied (4.50) to establish the second equality.

Thus, the forms F_s can be expressed by the forms D_s as follows

$$F_i(X) = -D_i(I_iX) + D_j(I_iX) + D_k(I_iX). \qquad (6.38)$$

Lemma 6.4.3. *Suppose* (M,η) *is a quaternionic contact manifold with constant qc-scalar curvature* $Scal = 16n(n+2)$. *Suppose* $\bar\eta = \frac{1}{2h}\eta$ *has vanishing* $[-1]$-*torsion component,* $\overline{T}^0 = 0$. *We have*

$$\nabla^*\left(\sum_{s=1}^{3} dh(\xi_s)F_s\right) = \sum_{s=1}^{3}\left[\nabla dh\,(I_se_a,\xi_s)F_s(I_se_a)\right]$$

$$+ h^{-1}\sum_{s=1}^{3}\left[dh(\xi_s)dh(I_se_a)D(e_a) + (n+2)\,dh(\xi_s)dh(I_se_a)\,A(e_a)\right].$$

Proof. We note that the vector $\sum_{s=1}^{3} dh(\xi_s)F_s$ is an $Sp(n)Sp(1)$ invariant vector. Hence, we may work with a qc-normal frame $\{e_1,\dots,e_{4n},\xi_1,\xi_2,\xi_3\}$. Since the scalar curvature is assumed to be constant we can apply Theorem 6.3.2, thus $\nabla^*T^0 = (n+2)A$. Turning to the divergence, we compute

$$\nabla^*\left(\sum_{s=1}^{3} dh(\xi_s)F_s\right) = -\sum_{s=1}^{3} h^{-1}\,dh(\xi_s)\,\nabla^*T^0(I_s\nabla h)$$

$$+ \sum_{s=1}^{3}\left[\nabla dh\,(e_a,\xi_s)F_s(e_a)\right] + \sum_{s=1}^{3}\left[h^{-2}\,dh(\xi_s)\,dh(e_a)T^0(e_a,I_se_b)\,dh(e_b)\right]$$

$$- \sum_{s=1}^{3}\left[h^{-1}\,dh(\xi_s)\,T^0(e_a,I_se_b)\,\nabla dh\,(e_a,e_b)\right]$$

$$= \sum_{s=1}^{3}\left[\nabla dh\,(e_a,\xi_s)F_s(e_a)\right] - \sum_{s=1}^{3} h^{-1}\,dh(\xi_s)\,\nabla^*T^0(I_s\nabla h)$$

$$+ \sum_{s=1}^{3}\left[h^{-1}\,dh(\xi_s)\,dh(I_se_a)\,D(e_a)\right]$$

$$= \sum_{s=1}^{3}\left[\nabla dh\,(e_a,\xi_s)F_s(e_a) + h^{-1}\,dh(\xi_s)\,dh(I_se_a)\,D(e_a)\right]$$

$$+ \sum_{s=1}^{3} h^{-1}(n+2)\,dh(\xi_s)\,dh(I_se_a)\,A(e_a), \qquad (6.39)$$

using the symmetry of T^0 in the next to last equality and the fact

$$T^0(e_a,I_1e_b)\,\nabla dh\,(e_a,e_b) = 0.$$

The latter can be seen, for example, by first using (5.12) and the formula for the symmetric part of ∇dh, the commutation relations (5.11), from which we have

$$T^0(e_a, I_1 e_b) \; \nabla dh \, (e_a, e_b)$$

$$= - \, h^{-1} \nabla dh_{[sym][-1]}(e_a, I_1 e_b) \Big[\nabla dh_{[sym]}(e_a, e_b) - \sum_{s=1}^{3} dh(\xi_s)\omega_s(e_a, e_b) \Big]$$

$$= \; h^{-1} \nabla dh_{[sym][-1]}(e_a, I_1 e_b) \nabla dh_{[sym][-1]}(e_a, e_b)$$

$$\quad - \; h^{-1} \nabla dh_{[sym][-1]}(e_a, I_1 e_b) \nabla dh_{[sym][3]}(e_a, e_b)$$

$$\quad + \; h^{-1} \nabla dh_{[sym][-1]}(e_a, I_1 e_b) \sum_{s=1}^{3} dh(\xi_s)\omega_s(e_a, e_b) = 0,$$

using the zero traces of the [-1]-component to justify the vanishing of the third term in the last equality. Switching to the basis $\{I_s e_a : a = 1, \dots, 4n\}$ in the first term of the right-hand-side of (6.39) completes the proof.   □

6.5   The divergence theorem in dimension seven

At this point we restrict our considerations to the 7-dimensional case, i.e. $n = 1$. Following is the main technical result. As mentioned in the introduction, a motivation to seek a divergence formula of this type based on the Riemannian and CR cases of the considered problem. The main difficulty is to find a suitable vector field with non-negative divergence containing the norm of the torsion. The fulfilment of this task was facilitated by the results of [94], which in particular showed that similarly to the CR case, but unlike the Riemannian case, we were not able to achieve a proof based purely on the Bianchi identities, see [[94], Theorem 4.8].

Theorem 6.5.1. *Suppose (M^7, η) is a quaternionic contact structure conformal to a positive 3-Sasakian structure $(M^7, \tilde{\eta})$, $\tilde{\eta} = \frac{1}{2h}\eta$. If $Scal_\eta = Scal_{\tilde{\eta}} = 16n(n+2)$, then with f given by*

$$f = \frac{1}{2} + h + \frac{1}{4}h^{-1}|\nabla h|^2,$$

the following identity holds

$$\nabla^* \Big(fD + \sum_{s=1}^{3} dh(\xi_s) F_s + 4 \sum_{s=1}^{3} dh(\xi_s) I_s A_s - \frac{10}{3} \sum_{s=1}^{3} dh(\xi_s) I_s A \Big)$$

$$= f|T^0|^2 + h \langle QV, V \rangle.$$

Here, Q is a positive semi-definite matrix and $V = (D_1, D_2, D_3, A_1, A_2, A_3)$ with A_s, D_s defined, correspondingly, in (6.20) and (6.35).

Proof. Using the formulas for the divergences of D, $\sum_{s=1}^{3} dh(\xi_s) F_s$, $\sum_{s=1}^{3} dh(\xi_s) I_s A_s$ and $\sum_{s=1}^{3} dh(\xi_s) I_s A$ given correspondingly in Lemmas 6.4.2, 6.4.3 and 6.4.1 we have the identity ($n = 1$ here)

$$\nabla^* \left(fD + \sum_{s=1}^{3} dh(\xi_s) F_s + 4 \sum_{s=1}^{3} dh(\xi_s) I_s A_s - \frac{10}{3} \sum_{s=1}^{3} dh(\xi_s) I_s A \right)$$

$$= \left(dh(e_a) - \frac{1}{4} h^{-2} dh(e_a) |\nabla h|^2 + \frac{1}{2} h^{-1} \nabla dh(e_a, \nabla h) \right) D(e_a)$$

$$+ f \left(|T^0|^2 - h^{-1} dh(e_a) D(e_a) - h^{-1}(n+2) dh(e_a) A(e_a) \right)$$

$$+ \sum_{s=1}^{3} \nabla dh(I_s e_a, \xi_s) F_s(I_s e_a)$$

$$+ h^{-1} \sum_{s=1}^{3} \left[dh(\xi_s) dh(I_s e_a) D(e_a) + (n+2) dh(\xi_s) dh(I_s e_a) A(e_a) \right]$$

$$+ 4 \sum_{s=1}^{3} \nabla dh(I_s e_a, \xi_s) A_s(e_a) - \frac{10}{3} \sum_{s=1}^{3} \nabla dh(I_s e_a, \xi_s) A(e_a)$$

$$= \left(dh(e_a) - \frac{1}{4} h^{-2} dh(e_a) |\nabla h|^2 + \frac{1}{2} h^{-1} \nabla dh(e_a, \nabla h) \right) \sum_{t=1}^{3} D_t(e_a)$$

$$+ f \left(|T^0|^2 - h^{-1} dh(e_a) \right) \left(\sum_{t=1}^{3} D_t(e_a) \right) - f h^{-1}(n+2) dh(e_a) \left(\sum_{t=1}^{3} A_t(e_a) \right)$$

$$+ \nabla dh(I_1 e_a, \xi_1) \left(D_1(e_a) - D_2(e_a) - D_3(e_a) \right)$$

$$+ \nabla dh(I_2 e_a, \xi_2) \left(-D_1(e_a) + D_2(e_a) - D_3(e_a) \right)$$

$$+ \nabla dh(I_3 e_a, \xi_3) \left(-D_1(e_a) - D_2(e_a) + D_3(e_a) \right)$$

$$+ h^{-1} \left(\sum_{s=1}^{3} dh(\xi_s) dh(I_s e_a) \right) \left(\sum_{t=1}^{3} D_t(e_a) \right)$$

$$+ h^{-1}(n+2) \left(\sum_{s=1}^{3} dh(\xi_s) dh(I_s e_a) \right) \left(\sum_{t=1}^{3} A_t(e_a) \right)$$

$$+ 4 \sum_{s=1}^{3} \nabla dh(I_s e_a, \xi_s) A_s(e_a) - \frac{10}{3} \left(\sum_{s=1}^{3} \nabla dh(I_s e_a, \xi_s) \right) \left(\sum_{t=1}^{3} A_t(e_a) \right),$$

$$(6.40)$$

where the last equality uses (6.38) to express the vectors F_s by D_s, and the expansions of the vectors A and D according to (6.20) and (6.36). Since the dimension of M is seven it follows $U = \bar{U} = [\nabla dh - 2h^{-1}dh \otimes dh]_{[3][0]} = 0$. This, together with the Yamabe equation (7.33), which when $n = 1$ becomes $\triangle h = 2 - 4h + 3h^{-1}|\nabla h|^2$, yield the formula, cf. (6.32),

$$\nabla dh\,(X, \nabla h) + \sum_{s=1}^{3} \nabla dh\,(I_s X, I_s \nabla h)$$

$$- \left(2 - 4h + 3h^{-1}|\nabla h|^2\right) dh(X) = 0. \quad (6.41)$$

From equations (6.35) and (6.31) we have

$$D_i(X) = h^{-2}\,dh(\xi_i)\,dh(I_i X) + \frac{1}{4}h^{-2}\left[\nabla dh\,(X, \nabla h) + \nabla dh\,(I_i X, I_i \nabla h)\right.$$
$$\left. - \nabla dh\,(I_j X, I_j \nabla h) - \nabla dh\,(I_k X, I_k \nabla h)\right].$$

Expressing the first term in (6.41) by the rest and substituting with the result in the above equations we come to

$$D_i(e_a) = \frac{1}{4}h^{-2}\left(2 - 4h + 3h^{-1}|\nabla h|^2\right) dh(e_a) + h^{-2}\,dh(\xi_i)\,dh(I_i e_a)$$
$$+ \frac{1}{2}h^{-2}\left[-\nabla dh\,(I_j e_a, I_j \nabla h) - \nabla dh\,(I_k e_a, I_k \nabla h)\right]. \quad (6.42)$$

At this point, by a purely algebraic calculation, using Lemma 6.3.3 and (6.42) we find:

$$\frac{22}{3}A_1 - \frac{2}{3}A_2 - \frac{2}{3}A_3 + \frac{11}{3}D_1 - \frac{1}{3}D_2 - \frac{1}{3}D_3$$

$$= -3h^{-1}\left(1 + \frac{1}{2}h^{-1}dh(e_a) + \frac{1}{4}h^{-2}|\nabla h|^2\right) dh(e_a)$$

$$+ 3h^{-2}\left(\sum_{s=1}^{3} dh(\xi_s)\,dh(I_s e_a)\right) + \frac{2}{3}h^{-1}\nabla dh(I_1 e_a, \xi_1)$$

$$- \frac{10}{3}h^{-1}\nabla dh(I_2 e_a, \xi_2) - \frac{10}{3}h^{-1}\nabla dh(I_3 e_a, \xi_3).$$

Similarly,

$$3\,A_1 - A_2 - A_3 + 2D_1 = \left(-2h^{-1} + \frac{1}{2}h^{-2} + h^{-3}|\nabla h|^2\right) dh(e_a)$$

$$- \frac{1}{2}h^{-2}\sum_{s=1}^{3}\nabla dh(I_s e_a, I_s \nabla h) + h^{-1}\,\nabla dh(I_1 e_a, \xi_1)$$

$$- h^{-1}\,\nabla dh(I_2 e_a, \xi_2) - h^{-1}\,\nabla dh(I_3 e_a, \xi_3) + h^{-2}\sum_{s=1}^{3} dh(\xi_s)\,dh(I_s e_a).$$

On the other hand, the coefficient of $A_1(e_a)$ in (6.40) is found to be, after setting $n = 1$,

$$h\left[-3\left(1 + \frac{1}{2}h^{-1} + \frac{1}{4}h^{-2}|\nabla h|^2\right)h^{-1}dh(e_a)\right.$$

$$+ 3h^{-2}\left(\sum_{s=1}^{3}dh(\xi_s)\,dh(I_s e_a)\right) + \frac{2}{3}h^{-1}\nabla dh(I_1 e_a, \xi_1)$$

$$\left. - \frac{10}{3}h^{-1}\nabla dh(I_2 e_a, \xi_2) - \frac{10}{3}h^{-1}\nabla dh(I_3 e_a, \xi_3)\right],$$

while the coefficient of $D_1(e_a)$ in (6.40) is

$$dh(e_a) - \frac{1}{4}h^{-2}dh(e_a)|\nabla h|^2 + \frac{1}{2}h^{-1}\,\nabla dh(e_a, \nabla h)$$

$$- f\,h^{-1}dh(e_a) + \nabla dh(I_1 e_a, \xi_1) - \nabla dh(I_2 e_a, \xi_2)$$

$$- \nabla dh(I_3 e_a, \xi_3)D_1(e_a) + h^{-1}\left(\sum_{s=1}^{3}dh(\xi_s)\,dh(I_s e_a)\right).$$

Substituting $\nabla dh(e_a, \nabla h)$ according to (6.41), i.e.,

$$\nabla dh\,(e_a, \nabla h) = -\sum_{s=1}^{3}\nabla dh\,(I_s e_a, I_s \nabla h) + \left(2 - 4h + 3h^{-1}|\nabla h|^2\right)dh(e_a)$$

and using the definition of f transforms the above formula into

$$dh(e_a) - \frac{1}{4}h^{-2}dh(e_a)|\nabla h|^2 - \left(\frac{1}{2} + h + \frac{1}{4}h^{-1}|\nabla h|^2\right)h^{-1}dh(e_a)$$

$$+ \frac{1}{2}h^{-1}\left(-\sum_{s=1}^{3}\nabla dh\,(I_s e_a, I_s \nabla h) + \left(2 - 4h + 3h^{-1}|\nabla h|^2\right)dh(e_a)\right)$$

$$+ \nabla dh(I_1 e_a, \xi_1) - \nabla dh(I_2 e_a, \xi_2) - \nabla dh(I_3 e_a, \xi_3)D_1(e_a)$$

$$+ h^{-1}\left(\sum_{s=1}^{3}dh(\xi_s)\,dh(I_s e_a)\right).$$

Simplifying the above expression shows that the coefficient of $D_1(e_a)$ in (6.40) is

$$\left(-2 + \frac{1}{2}h^{-1} + h^{-2}|\nabla h|^2\right)dh(e_a) - \frac{1}{2}h^{-1}\left(\sum_{s=1}^{3}\nabla dh\,(I_s e_a, I_s \nabla h)\right)$$

$$+ \nabla dh(I_1 e_a, \xi_1) - \nabla dh(I_2 e_a, \xi_2) - \nabla dh(I_3 e_a, \xi_3)$$

$$+ h^{-1}\left(\sum_{s=1}^{3}dh(\xi_s)\,dh(I_s e_a)\right).$$

Hence, we proved that the coefficient of $D_1(e_a)$ in (6.40) is

$$h\left(3\,A_1 - A_2 - A_3 + 2D_1\right)(e_a),$$

while those of $A_1(e_a)$ is

$$h\left(\frac{22}{3}\,A_1 - \frac{2}{3}\,A_2 - \frac{2}{3}\,A_3 + \frac{11}{3}\,D_1 - \frac{1}{3}\,D_2 - \frac{1}{3}\,D_3\right)(e_a).$$

A cyclic permutation gives the rest of the coefficients in (6.40). With this, the divergence (6.40) can be written in the form

$$\nabla^*\!\left(fD + \sum_{s=1}^{3} dh(\xi_s)\,F_s + 4\sum_{s=1}^{3} dh(\xi_s)I_sA_s - \frac{10}{3}\sum_{s=1}^{3} dh(\xi_s)\,I_sA\right)$$

$$= f|T^0|^2 + h\,\sigma_{1,2,3}\Big\{ g\left(D_1,\, 3A_1 - A_2 - A_3 + 2\,D_1\right)$$

$$+ g\left(A_1,\, \frac{22}{3}\,A_1 - \frac{2}{3}\,A_2 - \frac{2}{3}\,A_3 + \frac{11}{3}\,D_1 - \frac{1}{3}\,D_2 - \frac{1}{3}\,D_3\right)\Big\},$$

where $\sigma_{1,2,3}$ denotes the sum over all positive permutations of $(1,2,3)$. Let Q be equal to

$$Q := \begin{bmatrix} 2 & 0 & 0 & \dfrac{10}{3} & -\dfrac{2}{3} & -\dfrac{2}{3} \\[2mm] 0 & 2 & 0 & -\dfrac{2}{3} & \dfrac{10}{3} & -\dfrac{2}{3} \\[2mm] 0 & 0 & 2 & -\dfrac{2}{3} & -\dfrac{2}{3} & \dfrac{10}{3} \\[2mm] \dfrac{10}{3} & -\dfrac{2}{3} & -\dfrac{2}{3} & \dfrac{22}{3} & -\dfrac{2}{3} & -\dfrac{2}{3} \\[2mm] -\dfrac{2}{3} & \dfrac{10}{3} & -\dfrac{2}{3} & -\dfrac{2}{3} & \dfrac{22}{3} & -\dfrac{2}{3} \\[2mm] -\dfrac{2}{3} & -\dfrac{2}{3} & \dfrac{10}{3} & -\dfrac{2}{3} & -\dfrac{2}{3} & \dfrac{22}{3} \end{bmatrix}$$

so that

$$\nabla^*\!\left(fD + \sum_{s=1}^{3} dh(\xi_s)\,F_s + 4dh(\xi_s)\,I_sA_s - \frac{10}{3}\sum_{s=1}^{3} dh(\xi_s)\,I_sA\right)$$

$$= f\,|T^0|^2 + h\,\langle QV, V\rangle,$$

with $V = (D_1, D_2, D_3, A_1, A_2, A_3)$. It is not hard to see that the eigenvalues of Q are given by

$$\{0,\quad 0,\quad 2\,(2+\sqrt{2}),\quad 2\,(2-\sqrt{2}),\quad 10,\quad 10\},$$

which shows that Q is a non-negative matrix.   □

6.6   The qc Yamabe problem on the qc sphere and quaternionic Heisenberg group in dimension seven

In this section we give the proof of Theorem 6.1.2. We start with part a). Integrating the divergence formula of Theorem 6.5.1 it follows by Proposition 6.2.3 that the integral of the left-hand side is zero. Thus, the right-hand side vanishes as well, which shows that the quaternionic contact structure η has vanishing torsion endomorphism, i.e., it is also qc-Einstein according to Theorem 4.3.5.

Next we bring into consideration the 7-dimensional quaternionic Heisenberg group and the quaternionic Cayley transform as described in chapter 4 and 5. The quaternionic Heisenberg group of dimension seven is $G(\mathbb{H}) = \mathbb{H} \times \operatorname{Im}\mathbb{H}$. Let us identify the (seven dimensional) group $G(\mathbb{H})$ with the boundary Σ of a Siegel domain in $\mathbb{H} \times \mathbb{H}, \Sigma = \{(q',p') \in \mathbb{H} \times \mathbb{H} : \Re\, p' = |q'^2|\}$. Σ carries a natural group structure and the map $(q,\omega) \mapsto (q,|q|^2 - \omega) \in \Sigma$ is an isomorphism between $G(\mathbb{H})$ and Σ, see Section 5.2.1.

The standard contact form, written as a purely imaginary quaternion valued form, on $G(\mathbb{H})$ is given by $2\tilde{\Theta} = (d\omega - q \cdot d\bar{q} + dq \cdot \bar{q})$, where $\cdot$ denotes the quaternion multiplication. Since $dp = q \cdot d\bar{q} + dq \cdot \bar{q} - d\omega$, under the identification of $G(\mathbb{H})$ with Σ we also have $2\tilde{\Theta} = -dp' + 2dq' \cdot \bar{q}'$. Taking into account that $\tilde{\Theta}$ is purely imaginary, the last equation can be written also in the following form $4\tilde{\Theta} = (d\bar{p}' - dp') + 2dq' \cdot \bar{q}' - 2q' \cdot d\bar{q}'$.

Recall from Section 5.2.1 that the (quaternionic) Cayley transform is the map $\mathcal{C} : S \setminus \{(-1,0)\} \mapsto \Sigma$ from the sphere $S = \{(q,p) \in \mathbb{H} \times \mathbb{H} : |q|^2 + |p|^2 = 1\} \subset \mathbb{H} \times \mathbb{H}$ minus a point to the Heisenberg group $\Sigma = \{(q_1,p_1) \in \mathbb{H} \times \mathbb{H} : \Re\, p_1 = |q_1|^2\}$, with $\mathcal{C}$ defined by $(q_1,p_1) = \mathcal{C}\left((q,p)\right)$,

$$q_1 = (1+p)^{-1}\, q, \quad p_1 = (1+p)^{-1}\,(1-p). \qquad (6.43)$$

with an inverse $(q,p) = \mathcal{C}^{-1}\left((q_1,p_1)\right)$ given by

$$q = 2(1+p_1)^{-1}\, q_1, \quad p = (1-p_1)(1+p_1)^{-1}. \qquad (6.44)$$

By (5.29) the Cayley transform is a conformal quaternionic contact diffeomorphism between the quaternionic Heisenberg group with its standard quaternionic contact structure $\tilde{\Theta}$ and $S \setminus \{(-1,0)\}$ with the standard contact form on the sphere, $\tilde{\eta} = = dq \cdot \bar{q} + dp \cdot \bar{p} - q \cdot d\bar{q} - p \cdot d\bar{p}$. Hence, up to a constant multiplicative factor and a quaternionic contact automorphism the forms $\mathcal{C}_* \tilde{\eta}$ and $\tilde{\Theta}$ are conformal to each other. It follows that the same is true for $\mathcal{C}_* \eta$ and $\tilde{\Theta}$. In addition, $\tilde{\Theta}$ is qc-Einstein by definition, while η

and hence also $\mathcal{C}_*\eta$ are qc-Einstein as we observed at the beginning of the proof. According to Theorem 6.2.5, up to a multiplicative constant factor, the forms $\mathcal{C}_*\tilde{\eta}$ and $\mathcal{C}_*\eta$ are related by a translation or dilation on the Heisenberg group. Hence, we conclude that up to a multiplicative constant, η is obtained from $\tilde{\eta}$ by a conformal quaternionic contact automorphism which proves the first claim of Theorem 6.1.2. From the conformal properties of the Cayley transform and the existence Theorem 1.4.2 it follows that the minimum $\lambda(S^{4n+3})$ is achieved by a smooth 3-contact form, which due to the Yamabe equation is of constant qc-scalar curvature. This completes the proof of Theorem 6.1.2 a).

Next we prove part b) of Theorem 6.1.2. Let u be a positive entire solution of the Yamabe equation

$$\left(T_1^2 + X_1^2 + Y_1^2 + Z_1^2\right)u = -u^{3/2}. \tag{6.45}$$

As before, let $\tilde{\Theta}$ be the standard contact form on $G\,(\mathbb{H})$ identified with Σ. Using the inversion and the Kelvin transform on $G\,(\mathbb{H})$, see Sections 2.3.2 and 2.3.3, we can see that if $\Theta = \frac{1}{2h}\tilde{\Theta}$ has constant scalar curvature, then the Cayley transform lifts the qc structure defined by Θ to a qc structure of constant qc-scalar curvature on the sphere, which is conformal to the standard. Indeed, let us define two contact forms Θ_1 and Θ_2 on Σ setting

$$\Theta_1 = u^{4/(Q-2)}\tilde{\Theta}, \quad \text{and} \quad \Theta_2 = (\mathcal{K}u)^{4/(Q-2)}\,\frac{\bar{p}'}{|p'|}\,\tilde{\Theta}\,\frac{p'}{|p'|},$$

where u is as in (6.45), $\mathcal{K}u$ is its Kelvin transform, see (6.48) for the exact formula, and Q is the homogeneous dimension of the group. Notice that $\frac{\bar{p}'}{|p'|}\,\tilde{\Theta}\,\frac{p'}{|p'|}$ defines the same qc structure on the Heisenberg group as $\tilde{\Theta}$ and, in addition, $\mathcal{K}u$ is a smooth function on the whole group according to Theorem 2.3.7. Now, we shall invoke the Cayley transform in order to see that these two contact forms define a contact form on the sphere, which is conformal to the standard and has constant qc-scalar curvature. Let $P_1 = (-1,0)$ and $P_2 = (1,0)$ be correspondingly the 'south' and 'north' poles of the unit sphere $S = \{|q|^2 + |p|^2 = 1\}$. Let $\mathcal{C}_1$ and $\mathcal{C}_2$ be the corresponding Cayley transforms defined, respectively, on $S \setminus \{P_1\}$ and $S \setminus \{P_2\}$. Note that $\mathcal{C}_1$ was defined in (6.43), while $\mathcal{C}_2$ is given by $(q_2, p_2) = \mathcal{C}_2\left((q,p)\right)$,

$$q_2 = -(1-p)^{-1}\,q, \quad p_2 = (1-p)^{-1}\,(1+p). \tag{6.46}$$

In order that Θ_1 and Θ_2 define a contact form η on the sphere it is enough to see that

$$\Theta_1(p,q) = \Theta_2 \circ \mathcal{C}_2 \circ \mathcal{C}_1^{-1}(p,q), \quad \text{i.e.,} \quad \Theta_1 = (\mathcal{C}_2 \circ \mathcal{C}_1^{-1})^* \Theta_2. \tag{6.47}$$

A calculation shows that $C_2 \circ C_1^{-1} : \Sigma \to \Sigma$ is given by

$$q_2 = -p_1^{-1} q_1, \qquad p_2 = p_1^{-1},$$

or, equivalently, in the model $G(\mathbb{H})$

$$q_2 = -(|q_1|^2 - \omega_1)^{-1} q_1, \qquad \omega_2 = -\frac{\omega_1}{|q_1|^4 + |\omega_1|^2}.$$

Hence, $\sigma = C_2 \circ C_1^{-1}$ is an involution on the group and we have

$$C_{1*} \circ C_2^* \,\Theta = \frac{1}{|p_1|^2} \,\bar{\mu} \,\Theta \,\mu, \qquad \mu = \frac{p_1}{|p_1|},$$

which proves the identity (6.47). Using the properties of the Kelvin transform (2.37),

$$(\mathcal{K}u)(q', p') \stackrel{def}{=} |p'|^{-(Q-2)/2} u(\sigma(q', p')), \qquad (6.48)$$

proven in Theorem 2.3.7, we see that u and $\mathcal{K}u$ are solutions of the Yamabe equation (6.45). This implies that the contact form η has constant qc-scalar curvature, equal to $\frac{4(Q+2)}{Q-2}$.

Notice that η is conformal to the standard form $\tilde{\eta}$ and the arguments in the preceding proof imply then that η is qc-Einstein. A small calculation shows that this is equivalent to the fact that if we set

$$\bar{u} = 2^{10} \left[(1 + |q|^2)^2 + |\omega|^2 \right]^{-2}, \qquad (6.49)$$

then $\bar{u}$ satisfies the Yamabe equation (6.45) and all other nonnegative solutions of (6.45) in the space $\mathcal{D}^{1,2}$ are obtained from $\bar{u}$ by translations (1.28) and dilations (1.29) which in this case take the form

$$\tau_{(q_o, \omega_o)} \bar{u} (q, \omega) \stackrel{def}{=} \bar{u}(q_o + q, \omega + \omega_o), \qquad (6.50)$$

$$\bar{u}_\lambda (q) \stackrel{def}{=} \lambda^4 \bar{u}(\lambda q, \lambda^2 \omega), \qquad \lambda > 0. \qquad (6.51)$$

Thus, u which was defined in the beginning of the proof is given by equation (6.49) up to translations and dilations.

The proof of Theorem 6.1.2 is complete.

6.7 The qc Yamabe constant on the qc sphere and the best constant in the Folland-Stein embedding on the quaternionic Heisenberg group

In this section we shall prove Theorem 6.1.1. The proof relies on the realization made in [30] and used more recently in [70] that the "center of mass" idea of Szegö [158] and Hersch [89] can be used to find the sharp form of (logarithmic) Hardy-Littlewood-Sobolev type inequalities on the Heisenberg group. This method does not give all solutions of the qc Yamabe equation on the quaternionic contact sphere, but is enough to solve the Yamabe constant problem.

The conformal nature of the problem we consider is key to its solution. In this respect, even though the quaternionic contact (qc) Yamabe functional is involved, the qc scalar curvature is used in the proof without much geometric meaning. Rather, it is the conformal sub-Laplacian that plays a central role and the qc scalar curvature appears as a constant determined by the Cayley transform and the left-invariant sub-Laplacian on the quaternionic Heisenberg group. It should be possible to extend this method to a unified approach to all Iwasawa type groups using the results (and some further known facts) of Section 2.3.1.

The proof follows a sequence of lemmas, which we present next. We start with the determination of (the first) an eigenvalue and corresponding eigenfunctions of the horizontal sub-Laplacian on S^{4n+3}. In fact, we shall show that the restriction of every coordinate function is an eigenfunction.

Lemma 6.7.1. *If ζ is any of the (real) coordinate functions in $\mathbb{R}^{4n+4} = \mathbb{H}^n \times \mathbb{H}$, then*

$$\tilde{\triangle}\zeta = -\lambda_1\zeta, \quad \lambda_1 = \frac{\tilde{S}}{Q+2} = 2n, \tag{6.52}$$

where $\tilde{\triangle}$ is the horizontal sub-Laplacian (5.4) of the standard qc form $\tilde{\eta}$ on S^{4n+3} and $\tilde{S} = 8n(n+2)$ is its scalar curvature, cf. (5.31).

Proof. It is enough to prove the claim for one of the coordinate functions, say $\zeta = t_1$, since by virtue of its definition, the horizontal sub-Laplacian of the 3-Sasakian qc structure on the sphere is rotation invariant. Notice that ζ is quaternionic pluri-harmonic [Definition 6.7, [94]] since it is the real part of the anti-regular function $t_1 + ix_1 - jy_1 - kz_1$. Hence, its restriction to the 3-Sasakian sphere is the real part of an anti-CRF function. Therefore by [Corollary 6.24, [94]] it follows

$$\triangle\zeta = 4\lambda n$$

for the horizontal sub-Laplacian of the 3-Sasakian qc structure on the sphere, where $\lambda = \xi_1(x_1) - \xi_2(y_1) - \xi_3(z_1)$. Taking into account that the qc structure is 3-Sasakian it follows the Reeb vector fields are obtained from the outward pointing unit normal vector N as follows, $\xi_1 = iN$, $\xi_2 = jN$

and $\xi_3 = kN$, where for a point on the sphere we have $N(q) = q \in \mathbb{H}^{n+1}$. Therefore $\lambda = -t_1 = -\zeta$. Thus, for the horizontal sub-Laplacian of the 3-Sasakian qc sphere we have

$$\triangle \zeta = -4n\zeta,$$

where ζ is the restriction any of the coordinate functions of $\mathbb{R}^{4n+4} = \mathbb{H}^n \times \mathbb{H}$. Since the qc contact form $\tilde{\Theta}$ is twice the 3-Sasakian qc contact form on the sphere it follows $\tilde{\triangle}$ is $1/2$ of the 3-Sasakian sub-Laplacian. Thus

$$\tilde{\triangle} = -2n\zeta,$$

which shows $\lambda_1 = 2n = \frac{1}{2}(Q - 6) = \tilde{S}/(Q + 2)$. $\qquad\square$

Following is a key result allowing the ultimate solution of the considered problem. The idea follows Szegö and Hersch's center of mass method which was used on the Heisenberg group (2.9) in [30] and later in [70] in a manner which lead to the following lemma.

Lemma 6.7.2. *For every $v \in L^1(S^{4n+3})$ with $\int_{S^{4n+3}} v \, Vol_{\tilde{\eta}} = 1$ there is a quaternionic contact conformal transformation ψ such that*

$$\int_{S^{4n+3}} \psi v \, Vol_{\tilde{\eta}} = 0.$$

Proof. Let $P \in S^{4n+3}$ be any point of the quaternionic sphere and N be its antipodal point. Let us consider the local coordinate system near P defined by the Cayley transform $\mathcal{C}_N$ from N. As already used many times, $\mathcal{C}_N$ is a quaternionic contact conformal transformation between $S^{4n+3} \setminus N$ and the quaternionic Heisenberg group, cf. (5.29). Notice that in this coordinate system P is mapped to the identity of the group. For every r, $0 < r < 1$, let $\psi_{r,P}$ be the qc conformal transformation of the sphere, which in the fixed coordinate chart is given on the group by a dilation with center the identity by a factor δ_r. If we select a coordinate system in $\mathbb{R}^{4n+4} = \mathbb{H}^n \times \mathbb{H}$ so that $P = (1, 0)$ and $N = (-1, 0)$ and then apply the formulas for the Cayley transform the formula for $(q^*, p^*) = \psi_{r,P}(q, p)$ becomes

$$q^* = 2r \left(1 + r^2(1+p)^{-1}(1-p)\right)^{-1} (1 + p) q$$

$$p^* = \left(1 + r^2(1+p)^{-1}(1-p)\right)^{-1} \left(1 - r^2(1+p)^{-1}(1-p)\right), i.e,$$

We can define then the map $\Psi : B \to \bar{B}$, where B ($\bar{B}$) is the open (closed) unit ball in $\mathbb{R}^{4n+4}$, by the formula

$$\Psi(rP) = \int_{S^{4n+3}} \psi_{1-r,P} \, v \, Vol_{\tilde{\eta}}.$$

Notice that Ψ can be continuously extended to $\bar{B}$ since for any point P on the sphere, where $r = 1$, we have $\psi_{1-r,P}(Q) \to P$ when $r \to 1$. In particular, $\Psi = id$ on S^{4n+3}. Since the sphere is not a homotopy retract of the closed ball it follows that there are r and $P \in S^{4n+3}$ such that $\Psi(rP) = 0$, i.e., $\int_{S^{4n+3}} \psi_{1-r,P} \, v \, Vol_{\tilde{\eta}} = 0$. Thus, $\psi = \psi_{1-r,P}$ has the required property. $\qquad\square$

In the next step we prove that there is a minimizer of the Folland-Stein inequality which satisfies the zero center of mass condition. A number of well known invariance properties of the Yamabe functional will be exploited. For the rest of the chapter, given a qc form η and a function u we will denote by $\nabla^\eta u$ the horizontal gradient (5.3) of u.

Lemma 6.7.3. *Let v be a smooth positive function on the sphere with $\int_{S^{4n+3}} v^{2^*} Vol_{\tilde{\eta}} = 1$. There is a smooth positive function u such that $\int_{S^{4n+3}} \left(4\frac{Q+2}{Q-2} |\nabla^{\tilde{\eta}} u|^2 + \tilde{S} u^2 \right) Vol_{\tilde{\eta}} = \int_{S^{4n+3}} \left(4\frac{Q+2}{Q-2} |\nabla^{\tilde{\eta}} v|^2 + \tilde{S} v^2 \right) Vol_{\tilde{\eta}}$ and $\int_{S^{4n+3}} u^{2^*} Vol_{\tilde{\eta}} = 1$. In addition,*

$$\int_{S^{4n+3}} P \, u^{2^*}(P) \, Vol_{\tilde{\eta}} = 0, \qquad P \in \mathbb{R}^{4n+4} = \mathbb{H}^n \times \mathbb{H}. \tag{6.53}$$

In particular, the Yamabe constant

$$\lambda(S^{4n+3}, [\tilde{\eta}]) = \inf\left\{ \int_{S^{4n+3}} \left(4\frac{Q+2}{Q-2} |\nabla^{\tilde{\eta}} v|^2 + \tilde{S} v^2 \right) Vol_{\tilde{\eta}} : \right.$$
$$\left. \int_{S^{4n+3}} v^{2^*} Vol_{\tilde{\eta}} = 1, \ v > 0 \right\} \tag{6.54}$$

is achieved for a positive function u with a zero center of mass, i.e., for a function u satisfying (6.53).

Proof. Let $Vol_\eta = \eta_1 \wedge \eta_2 \wedge \eta_3 \wedge (\omega_1)^{2n}$ be the volume form associated to the qc contact form η. Thus if η is a qc structure on the sphere which is qc conformal to the standard qc structure $\tilde{\eta}$, $\eta = \phi^{4/(Q-2)} \tilde{\eta}$, then $Vol_\eta = \phi^{2^*} Vol_{\tilde{\eta}}$. This allows to put equation (6.3) in the form

$$\phi^{-1} v \, \mathcal{L}(\phi^{-1} v) \, Vol_\eta = v \tilde{\mathcal{L}}(v) \, Vol_{\tilde{\eta}}.$$

Therefore, if we take a positive function v on the sphere $\int_{S^{4n+3}} v^{2^*} Vol_{\tilde{\eta}} = 1$ and then consider the function

$$u = \phi^{-1}(v \circ \psi^{-1}), \tag{6.55}$$

where ψ is the qc conformal map of Lemma 6.7.2, $\eta \equiv (\psi^{-1})^* \tilde{\eta}$, and ϕ is the corresponding conformal factor of ψ, we can see that u achieves the claim of the Lemma. $\qquad\square$

We shall call a function u on the sphere a *well centered* function when (6.53) holds true. In the next step we show that a well centered minimizer has to be constant.

Lemma 6.7.4. *If u is a well centered local minimum of the problem* (6.54), *then $u \equiv const$.*

Proof. Let ζ be a smooth function on the sphere S^{4n+3} and

$$\mathcal{E}(v) \stackrel{def}{=} \int_{S^{4n+3}} \left(4\frac{Q+2}{Q-2} |\nabla^{\tilde{\eta}} v|^2 + \tilde{S} v^2 \right) Vol_{\tilde{\eta}}.$$

After applying the divergence formula, Proposition 6.2.3, we obtain the formula

$$\mathcal{E}(\zeta u) = \int_{S^{4n+3}} \zeta^2 \left(4\frac{Q+2}{Q-2} |\nabla^{\tilde{\eta}} u|^2 + \tilde{S} u^2 \right) Vol_{\tilde{\eta}}$$
$$- 4\frac{Q+2}{Q-2} \int_{S^{4n+3}} u^2 \zeta \tilde{\triangle} \zeta \, Vol_{\tilde{\eta}}. \quad (6.56)$$

Next, we let ζ be any of the coordinate functions in $\mathbb{H}^n \times \mathbb{H}$ in which case $\tilde{\triangle}\zeta = -\lambda_1 \zeta$ according to Lemma 6.7.1. It will be useful to introduce also the functional

$$N(v) = \left(\int_{S^{4n+3}} v^{2^*} Vol_{\tilde{\eta}} \right)^{2/2^*}$$

so that the Yamabe functional (6.2) can be written as $\Upsilon(v) = \mathcal{E}(v)/N(v)$ and the Yamabe constant is

$$\lambda(S^{4n+3}, [\tilde{\eta}]) = \inf\{\Upsilon(v) : v \in \overset{o}{D}{}^{1,2}(S^{4n+3})\}. \quad (6.57)$$

Computing the second variation $\delta^2\Upsilon(u)v = \frac{d^2}{dt^2}\Upsilon(u+tv)|_{t=0}$ of $\Upsilon(u)$ we see that the local minimum condition $\delta^2\Upsilon(u)v \geq 0$ implies

$$\mathcal{E}(v) - (2^* - 1)\mathcal{E}(u) \int_{S^{4n+3}} u^{2^*-2} v^2 \, Vol_{\tilde{\eta}} \geq 0$$

for any function v such that $\int_{S^{4n+3}} u^{2^*-1} v \, Vol_{\tilde{\eta}} = 0$. Therefore, for ζ being any of the coordinate functions in $\mathbb{H}^n \times \mathbb{H}$ we have

$$\mathcal{E}(\zeta u) - (2^* - 1)\mathcal{E}(u) \int_{S^{4n+3}} u^{2^*} \zeta^2 \, Vol_{\tilde{\eta}} \geq 0,$$

which after summation over all coordinate functions taking also into account (6.56) gives

$$\mathcal{E}(u) - (2^* - 1)\mathcal{E}(u) + 4\lambda_1(2^* - 1) \int_{S^{4n+3}} u^2 \, Vol_{\tilde{\eta}} \geq 0,$$

which implies, recall $2^* - 1 = (Q + 2)/(Q - 2)$,

$$0 \le 4(2^* - 1)(2^* - 2) \int_{S^{4n+3}} |\nabla^{\tilde{\eta}} u|^2 \, Vol_{\tilde{\eta}}$$

$$\le \left(4\lambda_1(2^* - 1) - (2^* - 2)\tilde{S} \right) \int_{S^{4n+3}} u^{2^*} \, Vol_{\tilde{\eta}}.$$

Thus, our task of showing that u is constant will be achieved once we see that

$$4\lambda_1(2^* - 1) - (2^* - 2)\tilde{S} \le 0, \quad \text{i.e,} \quad \lambda_1 \le S/(Q + 2). \tag{6.58}$$

By Lemma 6.52 we have actually equality $\lambda_1 = \tilde{S}/(Q + 2)$, which completes the proof. It is worth observing that inequality (6.58) can be written in the form

$$\lambda_1 \, a \le (2^* - 2)\tilde{S},$$

where a is the constant in front of the (sub-)Laplacian in the conformal (sub-)Laplacian, i.e., $a = 4\frac{Q+2}{Q-2}$ in our case. $\qquad \square$

After these preliminaries we turn to the proof of Theorem 6.1.1.

Proof of Theorem 6.1.1. Let F be a minimizer (local minimum) of the Yamabe functional $\mathcal{E}$ on $\boldsymbol{G}\,(\mathbb{H})$ and g the corresponding function on the sphere defined by

$$g = \mathcal{C}^*(F\Phi^{-1}). \tag{6.59}$$

By the conformality of the qc structures on the group and the sphere we have

$$Vol_{\Theta} = \Phi^{2^*} Vol_{\tilde{\Theta}}, \tag{6.60}$$

hence $F^{2^*} Vol_{\tilde{\Theta}} = f^{2^*} \phi^{-2^*} Vol_{\tilde{\eta}}$. This, together with the Yamabe equation implies that the Yamabe integral is preserved

$$\int_{\boldsymbol{G}\,(\mathbb{H})} |\nabla^{\tilde{\Theta}} F|^2 \, Vol_{\tilde{\Theta}} = \int_{S^{4n+3}} \left(|\nabla^{\tilde{\eta}} g|^2 + \frac{\tilde{S}}{a} g^2 \right) Vol_{\tilde{\eta}}, \tag{6.61}$$

where $a = 4(Q + 2)/(Q - 2)$. By Lemma 6.7.3 and (6.55) the function $g_0 = \phi^{-1}(g \circ \psi^{-1})$ will be well centered and a minimizer (local minimum) of the Yamabe functional Υ on S^{4n+3}. The latter claim uses also the fact that the map $v \mapsto u$ of equation (6.55) is one-to-one and onto on the space of smooth positive functions on the sphere. Now, from Lemma 6.7.4 we conclude that $g_o = const.$ Looking back at the corresponding functions on the group we see that

$$F_0 = \gamma \left[(1 + |q'|^2)^2 + |\omega'|^2 \right]^{-(Q-2)/4}$$

for some $\gamma = const. > 0$. Furthermore, the proof of Lemma 6.7.2 shows that F_0 is obtained from F by a translation (1.28) and dilation (1.29). Correspondingly, any positive minimizer (local minimum) of problem (3.85) is given up to dilation or translation by the function

$$F = \gamma \left[(1 + |q'|^2)^2 + |\omega'|^2 \right]^{-(Q-2)/4}, \qquad \gamma = const. > 0. \qquad (6.62)$$

Of course, translations (1.28) and dilations (1.29) do not change the value of $\mathcal{E}$. Incidentally, this shows that any local minimum of the Yamabe functional Υ on the sphere or the group has to be a global one.

We turn to the determination of the best constant. Let us define the constants

$$\Lambda_{\tilde{\Theta}} \stackrel{def}{=} \inf \left\{ \frac{\int_{\boldsymbol{G}(\mathbb{H})} |\nabla^{\tilde{\Theta}} v|^2 \, Vol_{\tilde{\Theta}}}{\left(\int_{\boldsymbol{G}(\mathbb{H})} |v|^{2^*} \, Vol_{\tilde{\Theta}} \right)^{2/2^*}} : v \in \overset{o}{\mathcal{D}}{}^{1,2}(\boldsymbol{G}), \, v > 0 \right\}$$

and

$$\Lambda \stackrel{def}{=} \inf \left\{ \frac{\int_{\boldsymbol{G}(\mathbb{H})} |\nabla^{\tilde{\Theta}} v|^2 \, dH}{\left(\int_{\boldsymbol{G}(\mathbb{H})} |v|^{2^*} \, dH \right)^{2/2^*}} : v \in \overset{o}{\mathcal{D}}{}^{1,2}(\boldsymbol{G}), \, v > 0 \right\},$$

where $\boldsymbol{G} = \boldsymbol{G}(\mathbb{H})$. Clearly, $\Lambda_{\tilde{\Theta}} = S_{\tilde{\Theta}}^{-2}$, where $S_{\tilde{\Theta}}$ is the best constant in the L^2 Folland-Stein inequality

$$\left(\int_{\boldsymbol{G}(\mathbb{H})} |u|^{2^*} \, Vol_{\tilde{\Theta}} \right)^{1/2^*} \leq S_{\tilde{\Theta}} \left(\int_{\boldsymbol{G}(\mathbb{H})} |\nabla^{\tilde{\Theta}} u|^2 \, Vol_{\tilde{\Theta}} \right)^{1/2}, \qquad (6.63)$$

while $\Lambda = S_2^{-2}$ is the best constant in the L^2 Folland-Stein inequality (1.1) (taken with respect to the Lebesgue measure !). Notice that we have the relation

$$Vol_{\tilde{\Theta}} = \kappa_{\tilde{\Theta}} \, dH, \qquad \kappa_{\tilde{\Theta}} = 2^{-3}(2n)! \qquad (6.64)$$

where dH is the Lebesgue measure in $\mathbb{R}^{4n+3}$, which is a Haar measure on the group. Thus, we have

$$\Lambda_{\tilde{\Theta}} = \kappa_{\tilde{\Theta}}^{1/(2n+3)} \Lambda. \qquad (6.65)$$

Furthermore, by Lemma 6.7.4 and equations (6.61) and (6.59) with $g = const$, we have

$$\Lambda_{\tilde{\Theta}} = \frac{1}{S_2^2} = \frac{\int_G |\nabla^{\tilde{\Theta}} F|^2 \, Vol_{\tilde{\Theta}}}{\left[\int_G |F|^{2^*} \, Vol_{\tilde{\Theta}}\right]^{2/2^*}} = \frac{\int_{S^{4n+3}} \left(|\nabla^{\tilde{\eta}} g|^2 + \frac{\tilde{s}}{a} g^2\right) Vol_{\tilde{\eta}}}{\left[\int_{S^{4n+3}} |g|^{2^*} \, Vol_{\tilde{\eta}}\right]^{2/2^*}}$$

$$- 2n(n+1)\left(\kappa_{\tilde{\eta}} \sigma_{4n+4}\right)^{1/(2n+3)} = 4n(n+1)\left[(2n)! \, \sigma_{4n+4}\right]^{1/(2n+3)}.$$

Here, $\sigma_{4n+4} = 2\pi^{2n+2}/\Gamma(2n+2) = 2\pi^{2n+2}/(2n+1)!$ is the volume of the unit sphere $S^{4n+3} \subset \mathbb{R}^{4n+4}$, see (6.67) below, and we also took into account Remark 6.2.4, which shows that $Vol_{\tilde{\eta}}$ gives $2^{2n+3}\left((2n)!\right)\omega_{4n+3}$ for the volume of S^{4n+3}. Alternatively, we could have used the function (6.62) and (3.91), which now reads

$$\int_{G(\mathbb{H})} \frac{1}{\left[(1+|x|^2)^2 + |y|^2\right]^{Q/2}} \, dH = \pi^{(4n+3)/2} \frac{\Gamma(\frac{4n+3}{2})}{\Gamma(4n+3)}, \qquad (6.66)$$

to compute $\Lambda_{\tilde{\Theta}}$. To compare the two values it is useful to recall some standard formulas involving the Euler's gamma function

$$\Gamma(n+1) = n!, \qquad \Gamma(z+n) = z(z+1)\ldots(z+n-1)\Gamma(z), \qquad n \in \mathbb{N},$$

$$\Gamma(2z) = 2^{2z-1}\,\pi^{-1/2}\,\Gamma(z)\,\Gamma\left(z + \frac{1}{2}\right) \qquad \text{– the Legendre formula,}$$

$$\sigma_{m+1} = 2\pi^{(m+1)/2}/\Gamma\left((m+1)/2\right) = \begin{cases} \frac{2^{(m+2)/2}\pi^{m/2}}{(m-1)!!}, & m\text{-even}, \\ \frac{2\pi^{(m+1)/2}}{(\frac{m-1}{2})!}, & m\text{-odd}. \end{cases}$$

$$(6.67)$$

Using either of these approaches, it follows that

$$S_{\tilde{\Theta}} = \frac{\left[(2n)! \, \sigma_{4n+4}\right]^{-1/(4n+6)}}{2\sqrt{n(n+1)}}. \qquad (6.68)$$

Furthermore, from (6.65) it follows the identity

$$\Lambda = 2n(n+1)\left(\frac{\kappa_{\tilde{\eta}}}{\kappa_{\tilde{\Theta}}}\sigma_{4n+4}\right)^{1/(2n+3)}$$

hence

$$S_2 = \frac{\left[2^{2n+6}\sigma_{4n+4}\right]^{-1/(4n+6)}}{\sqrt{2n(n+1)}},$$

which completes the proof of part a).

b) The Yamabe constant of the sphere is calculated immediately by taking a constant function in (6.57)

$$\lambda(S^{4n+3}, [\tilde{\eta}]) = a \, \Lambda_{\tilde{\Theta}}, \qquad a = 4\frac{Q+2}{Q-2} = 4\frac{n+2}{n+1}. \qquad (6.69)$$

This completes the proof of Theorem 6.1.1. □

Remark 6.7.5. The quaternionic case of the conjecture mentioned in Remark 3.5.3 follows. Indeed, the above computed value of Λ shows that $S_2^{ps} = S_2$ on the quaternionic Heisenberg group. Again, we should keep in mind that (3.92) uses an orthonormal basis which turns the group into a group of H-type. The standard basis (4.124) on $G\,(\mathbb{H})$ is not an Iwasawa basis, cf. also (2.10) and the paragraph above it. The two constants differ by a multiple of 4^{-3} (4^{-k} in the general case of a group of Iwasawa type with center of dimension k) - compare (6.66) and (3.91).

Chapter 7

CR manifolds - Cartan and Chern-Moser tensor and theorem

7.1 Introduction

In this section, based on [99], we give a new proof of the well known Chern-Moser theorem [45; 169] which states that a non-degenerate CR-hypersurface in $\mathbb{C}^{n+1}$, $n > 1$, is locally CR equivalent to a hyperquadric in $\mathbb{C}^{n+1}$ if and only if the Chern-Moser curvature vanishes. In dimension three we define a symmetric tensor and show that its vanishing is a sufficient condition a three dimensional CR-manifold to be locally CR equivalent to a hyperquadric in $\mathbb{C}^2$ thus giving a new proof of the Cartan theorem [40]. Our proof is based on the classical approach used by H.Weyl in Riemannian geometry [62]. The proof [99] is along the lines of the proof of Theorem 5.3.5 in the quaternionic contact geometry and it is instrumental to understand the whole proof of Theorem 5.3.5 in [97].

7.2 CR-manifolds and Tanaka-Webster connection

A CR manifold is a smooth manifold M of real dimension 2n+1, with a fixed n-dimensional complex subbbundle $\mathcal{H}$ of the complexified tangent bundle $\mathbb{C}TM$ satisfying $\mathcal{H} \cap \overline{\mathcal{H}} = 0$ and $[\mathcal{H}, \mathcal{H}] \subset \mathcal{H}$. If we let $H = Re\,\mathcal{H} \oplus \overline{\mathcal{H}}$, the real subbundle H is equipped with a formally integrable almost complex structure J. We assume that M is oriented and there exists a globally defined contact form θ such that $H = Ker\,\theta$. Recall that a 1-form θ is a contact form if the hermitian bilinear form

$$2g(X, Y) = -d\theta(JX, Y)$$

is non-degenerate. The vector field ζ dual to θ with respect to g satisfying $\zeta \lrcorner d\theta = 0$ is called the Reeb vector field. The almost complex structure J

is formally integrable in the sense that

$$([JX, Y] + [X, JY]) \in H$$

and the Nijenhuis tensor

$$N^J(X, Y) = [JX, JY] - [X, Y] - J[JX, Y] - J[X, JY] = 0.$$

A CR manifold (M, θ, g) with fixed contact form θ is called *a pseudohermitian manifold*. In this case the 2-form

$$d\theta_{|_H} := 2\Omega$$

is called the fundamental form. Note that the contact form is determined up to a conformal factor, i.e. $\bar{\theta} = \nu\theta$ for a positive smooth function ν, defines another pseudohermitian structure called pseudo-conformal to the original one.

Convention 7.2.1. In this chapter we use the conventions:

a) $X, Y, Z...$ will be a horizontal vector fields, i.e. $X, Y, Z... \in H$
b) $\{\epsilon_1, \dots, \epsilon_n, J\epsilon_1, \dots, J\epsilon_n\}$ denotes an adapted orthonormal basis (see for ex. [121]) of the horizontal space H.
c) The summation convention over repeated vectors from the basis $\{\epsilon_1, \dots, \epsilon_{2n}\}$ will be used.

The Tanaka-Webster connection [160; 169; 170] is the unique linear connection ∇^{cr} preserving a given pseudohermitian structure with torsion T^{cr} having the properties

$$\nabla^{cr}\zeta = \nabla^{cr}J = \nabla^{cr}\theta = \nabla^{cr}g = 0, \tag{7.1}$$

$$T^{cr}(X, Y) = d\theta(X, Y)\zeta = 2\Omega(X, Y)\zeta, \quad T^{cr}(\zeta, X) \in H, \tag{7.2}$$

$$g(T^{cr}(\zeta, X), Y) = g(T^{cr}(\zeta, Y), X) = -g(T^{cr}(\zeta, JX), JY). \tag{7.3}$$

It is well known that the endomorphism $T^{cr}(\zeta, .)$ is the obstruction a pseudohermitian manifold to be Sasakian. The symmetric endomorphism $T^{cr}_\zeta : H \longrightarrow H$ is denoted by A and it is call *the torsion of the pseudohermitian manifold*. The torsion A is completely trace-free satisfying

$$A(\epsilon_i, \epsilon_i) = A(\epsilon_i, J\epsilon_i) = 0, \quad A(X, Y) = A(Y, X) = -A(JX, JY). \tag{7.4}$$

Let R^{cr} be the curvature of the Tanaka-Webster connection. The pseudohermitian Ricci tensor r^{cr}, the psudohermitian scalar curvature s^{cr} and the pseudohermitian Ricci 2-form ρ^{cr} are defined by

$$r^{cr}(A, B) = R(\epsilon_i, A, B, \epsilon_i), \quad s^{cr} = r(\epsilon_i, \epsilon_i),$$

$$\rho^{cr}(A, B) = \frac{1}{2}R^{cr}(A, B, \epsilon_i, I\epsilon_i).$$

We summarize below the well known properties of the curvature R^{cr} of the Tanaka-Webster connection [169; 170; 120] using real expression, see also [60].

$$R^{cr}(X,Y,JZ,JV) = R^{cr}(X,Y,Z,V) = -R^{cr}(X,Y,V,Z),$$
$$R^{cr}(X,Y,Z,\zeta) = 0,$$

$$(7.5)$$

$$\frac{1}{2}\Big[R^{cr}(X,Y,Z,V) - R^{cr}(JX,JY,Z,V)\Big]$$
$$= -g(X,Z)A(Y,JV) - g(Y,V)A(X,JZ) + g(Y,Z)A(X,JV)$$
$$+ g(X,V)A(Y,JZ) - \Omega(X,Z)A(Y,V) - \Omega(Y,V)A(X,Z)$$
$$+ \Omega(Y,Z)A(X,V) + \Omega(X,V)A(Y,Z), \quad (7.6)$$

$$R^{cr}(\zeta,X,Y,Z) = (\nabla^{cr}_Y A)(Z,X) - (\nabla^{cr}_Z A)(Y,X), \tag{7.7}$$

$$r^{cr}(X,Y) = r^{cr}(Y,X),$$
$$r^{cr}(X,Y) - r^{cr}(JX,JY) = 4(n-1)A(X,JY),$$

$$(7.8)$$

$$2\rho^{cr}(X,JY) = -r^{cr}(X,Y) - r^{cr}(JX,JY) = R^{cr}(\epsilon_i,J\epsilon_i,X,JY), \tag{7.9}$$

$$2(\nabla^{cr}_{\epsilon_i} r^{cr})(\epsilon_i,X) = ds^{cr}(X).$$

$$(7.10)$$

7.3 The Cartan-Chern-Moser theorem

In this section we give a new proof of the well known Cartan-Chern-Moser theorem [40; 45; 169]. Our proof is based on the classical approach used by H. Weyl in Riemannian geometry [62].

Let v be a smooth function on a pseudohermitian manifold (M,θ,g). Let $\bar{\theta} = \frac{1}{2}e^{-2v}\theta$ be a pseudoconformal deformation of θ. We will denote the objects related to $\bar{\theta}$ by over-lining the same object corresponding to θ. Thus,

$$d\bar{\theta} = -e^{-2v}dv \wedge \theta + \frac{1}{2}e^{-2v}d\theta, \qquad \bar{g} = \frac{1}{2}e^{-2v}g. \tag{7.11}$$

The new Reeb vector field $\bar{\zeta}$ is

$$\bar{\zeta} = 2e^{2v}\zeta + 2e^{2v}J\nabla^{cr}v, \tag{7.12}$$

where $\nabla^{cr} v$ is the horizontal gradient, $g(\nabla^{cr} v, X) = dv(X)$. The horizontal sub-Laplacian and the norm of the horizontal gradient are defined respectively by

$$\triangle v \;=\; tr_H^g(\nabla^{cr} dv) \;=\; \nabla^{cr} dv(\epsilon_i, \epsilon_i), \quad |\nabla^{cr} v|^2 \;=\; dv(\epsilon_i)^2.$$

The connections ∇^{cr} and $\overline{\nabla^{cr}}$ are related by a (1,2) tensor $\mathbb{S}$,

$$\overline{\nabla^{cr}}_A B = \nabla^{cr}_A B + \mathbb{S}(A, B). \tag{7.13}$$

Suppose the contact structure is integrable. The conditions (7.2) and $\overline{\nabla^{cr}} \bar{g} = 0$ determine $g(\mathbb{S}(X, Y), Z)$ due to the equality

$$g(\mathbb{S}(X,Y),Z) = -dv(X)g(Y,Z) + dv(JX)\Omega(Y,Z) - dv(Y)g(Z,X)$$
$$- dv(JY)\Omega(Z,X) + dv(Z)g(X,Y) - dv(JZ)\Omega(X,Y). \tag{7.14}$$

We obtain after some calculations using (7.12) that

$$\bar{A}(X,Y) - 2e^{2u} A(X,Y) - g(\mathbb{S}(\bar{\zeta},X),Y)$$
$$= -2e^{2v}\nabla^{cr} dv(X,JY) - 4e^{2v} dv(X)dv(JY). \tag{7.15}$$

From (7.15) and (7.3) we find

$$g(\mathbb{S}(\bar{\zeta},X),Y) + g(\mathbb{S}(\bar{\zeta},JX)JY)$$
$$= 2e^{2v}\Big[\nabla^{cr} dv(X,JY) - \nabla^{cr} dv(JX,Y)\Big]$$
$$+ 4e^{2v}\Big[dv(X)dv(JY) - dv(JX)dv(Y)\Big]. \tag{7.16}$$

The conditions $\overline{\nabla^{cr}} J = \nabla^{cr} J = 0$ yield $g(\mathbb{S}(\bar{\zeta},X),Y) = g(\mathbb{S}(\bar{\zeta},JX)JY)$. Substitute the latter into (7.15) and (7.16), use (7.12) and (7.14) to get

$$g(\mathbb{S}(\zeta,X),Y) = \frac{1}{2}\Big[\nabla^{cr} dv(X,JY) - \nabla^{cr} dv(JX,Y)\Big]$$
$$- dv(X)dv(JY) + dv(JX)du(Y) + |\nabla^{cr} v|^2\Omega(X,Y) \tag{7.17}$$

and for the pseudohermitian torsion we obtain [121]

$$\bar{A}(X,Y) = 2e^{2v}\Big[A(X,Y) - dv(X)dv(JY) - dv(JX)dv(Y)\Big]$$
$$- e^{2v}\Big[\nabla^{cr} dv(X,JY) + \nabla^{cr} dv(JX,Y)\Big]. \tag{7.18}$$

The identity $d^2 = 0$ together with (7.2) yields

$$\nabla^{cr} dv(X,Y) = [\nabla^{cr} dv]_{[sym]}(X,Y) - dv(\zeta)\Omega(X,Y). \tag{7.19}$$

We consider the symmetric (0,2) tensor C defined on H by the equality

$$C(X,Y) = A(JX,Y)$$

$$- \frac{1}{2(n+2)}\rho^{cr}(X,JY) - \frac{s^{cr}}{8(n+1)(n+2)}g(X,Y) \quad (7.20)$$

and define the (0,4) tensor CW on H by

$$g(CW(X,Y)Z,V) = g(R(X,Y)Z,V) + g(X,Z)C(Y,V) + g(Y,V)C(X,Z)$$

$$-g(Y,Z)C(X,V) - g(X,V)C(Y,Z) - \Omega(X,Z)C(Y,JV) - \Omega(Y,V)C(X,JZ)$$

$$+\Omega(Y,Z)C(X,JV) + \Omega(X,V)C(Y,JZ) - \Omega(X,Y)\Big[C(Z,JV) - C(JZ,V)\Big]$$

$$- \Omega(Z,V)\Big[C(X,JY) - C(JX,Y)\Big]. \quad (7.21)$$

Take the corresponding traces in (7.21) keeping in mind (7.20) to verify that the tensor CW is completely trace-free, i.e. $r^{cr}(CW) = \rho^{cr}(CW) = 0$.

Compare (7.21) with (7.7) to obtain the following

Proposition 7.3.1. *The (2,0)+(0,2)-part of the tensor CW vanishes identically,*

$$CW(X,Y,Z,V) - CW(JX,JY,Z,V) = 0.$$

The (1,1)-part of CW is precisely the Chern-Moser tensor S defined in [45] and it is determined completely by the Ricci 2-form as follows

$$S(X,Y,Z,V) = CW_{1,1}(X,Y,Z,V)$$

$$= \frac{1}{2}\Big[CW(X,Y,Z,V) + CW(JX,JY,Z,V)\Big]$$

$$= \frac{1}{2}\Big[R(X,Y,Z,V) + R(JX,JY,Z,V)\Big]$$

$$- \frac{s^{cr}}{4(n+1)(n+2)}\Big[g(X,Z)g(Y,V) - g(Y,Z)g(X,V)$$

$$+ \Omega(X,Z)\Omega(Y,V) - \Omega(Y,Z)\Omega(X,V) + 2\Omega(X,Y)\Omega_s(Z,V)\Big]$$

$$- \frac{1}{2(n+2)}\Big[g(X,Z)\rho^{cr}(Y,JV) - g(Y,Z)\rho^{cr}(X,JV)\Big]$$

$$- \frac{1}{2(n+2)}\Big[g(Y,V)\rho^{cr}(X,JZ) - g(X,V)\rho^{cr}(Y,JZ)\Big]$$

$$- \frac{1}{2(n+2)}\Big[\Omega(X,Z)\rho^{cr}(Y,V) - \Omega(Y,Z)\rho^{cr}(X,V)\Big]$$

$$- \frac{1}{2(n+2)}\Big[\Omega(Y,V)\rho^{cr}(X,Z) - \Omega(X,V)\rho^{cr}(Y,Z)\Big]$$

$$- \frac{1}{n+2}\Big[\Omega(X,Y)\rho^{cr}(Z,V) + \Omega(Z,V)\rho^{cr}(X,Y)\Big]. \quad (7.22)$$

For $n = 1$ the tensor S vanishes identically and we consider the following $(0,2)$ tensor defined on H by

$$F^{car}(X, Y) = (\nabla^{cr} ds^{cr})(X, JY) + (\nabla^{cr} ds^{cr})(Y, JX)$$
$$+ 16((\nabla^{cr})^2_{X\epsilon_a} A)(Y, \epsilon_a) + 16((\nabla^{cr})^2_{Y\epsilon_a} A)(X, \epsilon_a) + 36s^{cr} A(X, Y)$$
$$+ 48((\nabla^{cr})^2_{\epsilon_a J\epsilon_a} A)(X, JY) + 3g(X, Y)(\nabla^{cr} ds^{cr})(\epsilon_a, J\epsilon_a). \quad (7.23)$$

We call the tensor F^{car} the *Cartan tensor* in view of Theorem 7.3.5 below. The relevance of the tensors CW and F^{car} become clear from the next result and Theorem 7.3.5 below proved by Chern and Moser [45] and Cartan [40] in a completely different way using the Cartan method of equivalences

Theorem 7.3.2. *The tensor CW is a pseudohermitian invariant, i.e. if v is a smooth function and $2\bar{\theta} = e^{-2v}\theta$, then $2e^{2v}CW_{\bar{\theta}} = CW_\theta$.*

Proof. By a straightforward computations using (7.14) and (7.17), we obtain the formula

$$2e^{2v}g(\bar{R}(X, Y)Z, V) - g(R(X, Y)Z, V) = -g(Z, V)\Big[N(X, Y) - N(Y, X)\Big]$$
$$- g(X, Z)N(Y, V) - g(Y, V)N(X, Z) + g(Y, Z)N(X, V) + g(X, V)N(Y, Z)$$
$$+ \Omega(X, Z)N(Y, JV) + \Omega(Y, V)N(X, JZ) - \Omega(Y, Z)N(X, JV)$$
$$- \Omega(X, V)N(Y, JZ) + \Omega(X, Y)\Big[N(Z, JV) - N(JZ, V)\Big]$$
$$+ \Omega(Z, V)\Big[N(X, JY) - N(Y, JX)\Big], \quad (7.24)$$

where the $(0,2)$ tensor N is given by

$$N(X, Y) = \nabla^{cr} dv(X, Y)$$
$$+ dv(X)dv(Y) - dv(JX)dv(JY) - \frac{1}{2}g(X, Y)|\nabla^{cr} v|^2. \quad (7.25)$$

We denote $trN = N(e_a, e_a)$ the trace of the tensor N. Using (7.25) and (7.19), we obtain

$$trN = \triangle v - n|dv|^2,$$
$$N(X, Y) - N(JX, JY) = N(Y, X) - N(JY, JX), \quad (7.26)$$

$$N(\epsilon_i, J\epsilon_i) = -2ndv(\zeta),$$
$$N(\epsilon_i, J\epsilon_i)\Omega(X, Y) = n\Big[N(X, Y) - N(Y, X)\Big]. \quad (7.27)$$

Taking the traces in (7.24) and using (7.25), (7.26) and (7.27) we obtain

$$\overline{r}^{cr}(X,Y) - r^{cr}(X,Y) = (n+1)N(X,Y) + nN(Y,X)$$
$$+ N(JX,JY) + 2N(JY,JX) + trN\,g(X,Y), \qquad (7.28)$$
$$e^{-2v}\overline{s}^{cr} - 2s^{cr} = 8(n+1)trN.$$

Equation (7.28) together with (7.9) and (7.8) imply

$$\overline{r}^{cr}(X,Y) - r^{cr}(X,Y) + \overline{r}^{cr}(JX,JY) - r^{cr}(JX,JY)$$
$$= -2\Big[\overline{\rho^{cr}}(X,JY) - \rho^{cr}(X,JY)\Big] = 2(trN)\,g(X,Y)$$
$$+ (n+2)\Big[N(X,Y) + N(Y,X) + N(JX,JY) + N(JY,JX)\Big], \quad (7.29)$$

$$\overline{r}^{cr}(X,Y) - r^{cr}(X,Y) - \overline{r}^{cr}(JX,JY) + r^{cr}(JX,JY)$$
$$= 4(n-1)\Big[\overline{A}(X,JY) - A(X,JY)\Big]$$
$$= (n-1)\Big[N(X,Y) + N(Y,X) - N(JX,JY) - N(JY,JX)\Big]. \quad (7.30)$$

The equalities (7.28), (7.29) and (7.30) yield

$$N_{[sym]}(X,Y)$$
$$= -\frac{1}{2(n+2)}\overline{\rho}^{cr}(X,JY) - \frac{\overline{s}^{cr}}{8(n+1)(n+2)}\overline{g}(X,Y) + \overline{A}(JX,Y)$$
$$- \Big[-\frac{1}{2(n+2)}\rho^{cr}(X,JY) - \frac{s^{cr}}{8(n+1)(n+2)}g(X,Y) + A(JX,Y)\Big]$$
$$= \overline{C}(X,Y) - C(X,Y). \quad (7.31)$$

We obtain from (7.25) and (7.19) that

$$N(X,Y) = N_{[sym]}(X,Y) - dv(\zeta)\omega(X,Y). \qquad (7.32)$$

Substituting (7.31) in (7.32), and then inserting the obtained equality in (7.24), invoking also (7.26), completes the proof of Theorem 7.3.2. □

Remark 7.3.3. If we substitute (7.25), (7.26) and (7.27) in (7.28) one recovers the transformation formulas of the pseudo hermitian Ricci tensor and the pseudo hermitian scalar found in [121]

$$\overline{s^{cr}} = 2e^{2v}s^{cr} - 8n(n+1)e^{2v}|\nabla^{cr}v|^2 + 8(n+1)e^{2v}\triangle v. \qquad (7.33)$$

The result due to Chern and Moser [45] follows from Theorem 7.3.2 and Proposition 7.3.1.

Theorem 7.3.4. *[45] The Chern-Moser tensor S is a pseudo-conformal invariant.*

The rest of the section is devoted to a new proof of the next theorem due to Chern-Moser [45] and Webster [169] in dimension bigger than three and due to Cartan [40] in dimension three.

Theorem 7.3.5. *[40; 45; 169] Let (M, θ, g) be a $2n+1$-dimensional non-degenerate pseudo-hermitian manifold.*

i) [45; 169] If $n > 1$ then (M, θ, g) is locally pseudoconformal equivalent to a hyperquadric in $\mathbb{C}^{n+1}$ if and only if the Chern-Moser tensor vanishes, $S = 0$;

ii) [40] If $n = 1$ then (M, θ, g) is locally pseudoconformal equivalent to a hyperquadric in $\mathbb{C}^2$ if and only the tensor F^{car} vanishes, $F^{car} = 0$.

Proof. It is well known that a pseudohermitian manifold with flat Tanaka-Webster connection is locally isomorphic to the (complex) Heisenberg group. For $n > 1$ the vanishing of the horizontal part of the Tanaka-Webster connection implies the vanishing of the whole curvature, as follows from (7.8) and (7.7). If $n = 1$, in addition to the vanishing of the horizontal part of the curvature one needs also the vanishing of the pseudohermitian torsion to have zero curvature.

On the other hand, the complex Cayley transform is a pseudo-conformal equivalence between the Heisenberg group with its flat pseudo-hermitian structure and a hypersphere $h_{\alpha\bar{\beta}}Z^\alpha Z^{\bar{\beta}} + W\bar{W} = 1$ in $\mathbb{C}^{n+1}$ [[45], p.223]. We shall show that the condition $S = 0$ is a sufficient condition a given pseudohermitian manifold to be locally pseudoconformally flat on H provided the dimension is bigger than three. In dimension three S vanishes identically and the sufficient condition is the vanishing of the tensor F^{car} which also implies the vanishing of the pseudohermitian torsion.

Suppose $S = 0$. Then $CW = 0$ due to Proposition 7.3.1. We shall show that in this case there (locally) exists a smooth function v which sends the pseudo-hermitian structure to the flat one by a pseudoconformal transformation.

We consider the following system of differential equations with respect to an unknown function v:

$$\nabla^{cr}dv(X,Y) = -C(X,Y) - dv(X)dv(Y)$$
$$+ dv(JX)dv(JY) + \frac{1}{2}g(X,Y)|\nabla^{cr}v|^2 - dv(\zeta)\Omega(X,Y), \quad (7.34)$$

$$\nabla^{cr}dv(X,\zeta) = -\mathbb{D}(X,\zeta)$$
$$- C(X, J\nabla^{cr}v) + \frac{1}{2}dv(JX)|\nabla^{cr}v|^2 - dv(X)dv(\zeta), \quad (7.35)$$

$$\nabla^{cr} dv(\zeta, \zeta) = -\mathbb{D}(\zeta, \zeta) - \mathbb{D}(J\nabla^{cr}v, \zeta) + \frac{1}{4}|\nabla^{cr}v|^4 - (dv(\zeta))^2, \quad (7.36)$$

where $\mathbb{D}(X, \zeta)$ and $\mathbb{D}(\zeta, \zeta)$ do not depend on the function v and are determined by

$$\begin{aligned}(\nabla^{cr}_{\epsilon_i} C)(J\epsilon_i, JX) &= -(2n+1)\mathbb{D}(JX, \zeta) \\ -(\nabla^{cr}_X tr\, C) + (\nabla^{cr}_{\epsilon_i} C)(\epsilon_i, X) &= 3\mathbb{D}(JX, \zeta),\end{aligned} \quad (7.37)$$

$$\mathbb{D}(\zeta, \zeta) = -\frac{1}{2n}[(\nabla^{cr}_{\epsilon_i}\mathbb{D})(J\epsilon_i, \zeta) + C(\epsilon_i, J\epsilon_j)C(J\epsilon_i, \epsilon_j)]. \quad (7.38)$$

The consistences of the first and second equality in (7.37) is precisely equivalent to (7.10).

To prove Theorem 7.3.5 it is sufficient to show the existence of a local smooth solution to (7.34) because of (7.18) and the proof of Theorem 7.3.2. However, if equation (7.34) has a smooth solution then (7.35) and (7.36) appear as necessary conditions, so we considered the complete system (7.34)-(7.36) and reduced the question to showing that this system has (locally) a smooth solution.

The integrability conditions for the overdetermined system (7.34)-(7.36) are furnished by the Ricci identities,

$$\begin{aligned}\nabla^{cr} dv(A, B, C) - \nabla^{cr} dv(B, A, C) = \\ - R^{cr}(A, B, C, \nabla^{cr}v) - \nabla^{cr} dv((T^{cr}(A, B), C). \quad (7.39)\end{aligned}$$

We consider all possible cases.

Case 1: $\left[Z, X, Y \in H\right]$. The equation (7.39) on H has the following form

$$\begin{aligned}\nabla^{cr} dv(Z, X, Y) - \nabla^{cr} dv(X, Z, Y) \\ = -R^{cr}(Z, X, Y, \nabla^{cr}v) - 2\Omega(Z, X)\nabla^{cr} dv(\zeta, Y), \quad (7.40)\end{aligned}$$

where we have used (7.2).

When we take the covariant derivative of (7.34) along $Z \in H$, substitute in the obtained equality (7.34) and (7.35), then anticommute the covariant derivatives, substitute into (7.40), and use (7.21) with $S = 0 = CW$ we see that the integrability condition here is

$$\begin{aligned}(\nabla^{cr}_Z C)(X, Y) - (\nabla^{cr}_X C)(Z, Y) \\ = -\Omega(Z, Y)\mathbb{D}(X, \zeta) + \Omega(X, Y)\mathbb{D}(Z, \zeta) - 2\Omega(Z, X)\mathbb{D}(Y, \zeta). \quad (7.41)\end{aligned}$$

Lemma 7.3.6. *If the dimension of M is bigger than three and $S = 0$, then (7.41) holds.*

Proof. Using (7.2), the second Bianchi identity (4.113) gives

$$\sum_{(X,Y,Z)} \left[(\nabla_X^{cr} R^{cr})(Y,Z,V,W) + 2\Omega(X,Y)R^{cr}(\zeta,Z,V,W) \right] = 0. \quad (7.42)$$

Trace in (7.42) yields

$$(\nabla_{\epsilon_i}^{cr} R^{cr})(X,Y,Z,\epsilon_i) - (\nabla_X^{cr} r^{cr})(Y,Z) + (\nabla_Y^{cr} r^{cr})(X,Z) \quad (7.43)$$
$$-2R^{cr}(\zeta,Y,Z,JX) + 2R^{cr}(\zeta,X,Z,JY) - 2\Omega(X,Y)r^{cr}(\zeta,Z) = 0,$$

$$(\nabla_X^{cr} \rho^{cr})(Y,Z) + (\nabla_Y^{cr} \rho^{cr})(Z,X) + (\nabla_Z^{cr} \rho^{cr})(X,Y) \quad (7.44)$$
$$+2\Omega(X,Y)\rho^{cr}(\zeta,Z) + 2\Omega(Y,Z)\rho^{cr}(\zeta,X) + 2\Omega(Z,X)\rho^{cr}(\zeta,Y) = 0,$$

$$(\nabla_X^{cr} \rho^{cr})(Y,Z) + (\nabla_{e_a}^{cr} R^{cr})(J\epsilon_i,X,Y,Z,) \quad (7.45)$$
$$+2(n-1)R^{cr}(\zeta,X,Y,Z) = 0.$$

We use $S = 0$ and (7.20) to express r^{cr}, ρ^{cr} and A in terms of C and $tr\, C$, namely

$$\rho^{cr}(X,Y)$$
$$= (n+2)C(X,JY) - (n+2)C(JX,Y) - (tr\,C)\Omega(X,Y), \quad (7.46)$$

$$r^{cr}(X,Y) = (2n+1)C(X,Y) + 3C(JX,JY) + (tr\,C)g(X,Y), \quad (7.47)$$

$$A(JX,Y) = \frac{1}{2}[C(X,Y) - C(JX,JY)]. \quad (7.48)$$

Inserting (7.21) and (7.7) in (7.43), and then using (7.47), (7.48) and (7.37) we come after some standard calculations to the following identity

$$-3g(Z,X)\mathbb{D}(JY,\zeta) + 3g(Z,Y)\mathbb{D}(JX,\zeta)$$
$$- (2n+1)\Omega(X,Z)\mathbb{D}(Y,\zeta) + (2n+1)\Omega(Y,Z)\mathbb{D}(X,\zeta)$$
$$- 2(2n+1)\Omega(X,Y)\mathbb{D}(Z,\zeta) - 2n[(\nabla_X^{cr}C)(Y,Z) - (\nabla_Y^{cr}C)(X,Z)]$$
$$+ [(\nabla_{JZ}^{cr}C)(X,JY) - (\nabla_{JZ}^{cr}C)(JX,Y)]$$
$$- [(\nabla_{JX}^{cr}C)(JY,Z) - (\nabla_{JY}^{cr}C)(JX,Z)]$$
$$- 2[(\nabla_{JX}^{cr}C)(Y,JZ) - (\nabla_{JY}^{cr}C)(X,JZ)]$$
$$- 3[(\nabla_X^{cr}C)(JY,JZ) - (\nabla_Y^{cr}C)(JX,JZ)] = 0. \quad (7.49)$$

A substitution of (7.21) and (7.7) in (7.44) together with (7.46) give

$$(n+2)[(\nabla_X^{cr}C)(Y,JZ) - (\nabla_Y^{cr}C)(X,JZ)]$$
$$- (n+2)[(\nabla_X^{cr}C)(JY,Z) - (\nabla_Y^{cr}C)(JX,Z)]$$
$$+ (n+2)[(\nabla_Z^{cr}C)(X,JY) - (\nabla_Z^{cr}C)(JX,Y)] + 2(n+2)[\Omega(X,Y)\mathbb{D}(JZ,\zeta)$$
$$+ \Omega(Y,Z)\mathbb{D}(JX,\zeta) + \Omega(Z,X)\mathbb{D}(JY,\zeta)] = 0. \quad (7.50)$$

Taking JZ instead of Z in (7.50), and then evaluating at JX and JY, correspondingly, instead of X and Y into the obtained result we derive after a subtraction of the thus achieved equalities the following identity

$$- [(\nabla^{cr}_X C)(Y, Z) - (\nabla^{cr}_Y C)(X, Z)]$$
$$- [(\nabla^{cr}_X C)(JY, JZ) - (\nabla^{cr}_Y C)(JX, JZ)]$$
$$+ [(\nabla^{cr}_{JX} C)(JY, Z) - (\nabla^{cr}_{JY} C)(JX, Z)]$$
$$- [(\nabla^{cr}_{JX} C)(Y, JZ) - (\nabla^{cr}_{JY} C)(X, JZ)] +$$
$$+ 2g(Y, Z)\mathbb{D}(JX, \zeta) - 2g(Z, X)\mathbb{D}(JY, \zeta)$$
$$+ 2\Omega(Y, Z)\mathbb{D}(X, \zeta) - 2\Omega(X, Z)\mathbb{D}(Y, \zeta) = 0. \quad (7.51)$$

Now we substitute (7.21), (7.7) in (7.45) using (7.46), (7.48), then replace Y and Z, respectively, with JY and JZ in the obtained equality and then take the difference of both equations to obtain

$$(n - 1)[(\nabla^{cr}_{JZ} C)(X, JY) - (\nabla^{cr}_{JY} C)(X, JZ)]$$
$$+ (n - 1)[(\nabla^{cr}_{JZ} C)(JX, Y) - (\nabla^{cr}_{JY} C)(JX, Z)]$$
$$- (n - 1)[(\nabla^{cr}_Z C)(X, Y) - (\nabla^{cr}_Y C)(X, Z)]$$
$$+ (n - 1)[(\nabla^{cr}_Z C)(JX, JY) - (\nabla^{cr}_Y C)(JX, JZ)] = 0. \quad (7.52)$$

Substituting X by Z, and Z by X in (7.52) and then summing the obtained equalities and (7.51) yields

$$- [(\nabla^{cr}_X C)(Y, Z) - (\nabla^{cr}_Y C)(X, Z)]$$
$$+ [(\nabla^{cr}_{JX} C)(JY, Z) - (\nabla^{cr}_{JY} C)(JX, Z)]$$
$$- g(X, Z)\mathbb{D}(JY, \zeta) + g(Y, Z)\mathbb{D}(JX, \zeta)$$
$$- \Omega(X, Z)\mathbb{D}(Y, \zeta) + \Omega(Y, Z)\mathbb{D}(X, \zeta) = 0. \quad (7.53)$$

The cyclic sum in (7.53) gives

$$[(\nabla^{cr}_{JZ} C)(X, JY) - (\nabla^{cr}_{JZ} C)(JX, Y)]$$
$$= [(\nabla^{cr}_{JX} C)(JY, Z) - (\nabla^{cr}_{JY} C)(JX, Z)]$$
$$- [(\nabla^{cr}_{JX} C)(Y, JZ) - (\nabla^{cr}_{JY} C)(X, JZ)]$$
$$+ 2\Omega(Z, X)\mathbb{D}(Y, \zeta) + 2\Omega(Y, Z)\mathbb{D}(X, \zeta) + 2\Omega(X, Y)\mathbb{D}(Z, \zeta). \quad (7.54)$$

Now, identity (7.41) follows from (7.49), (7.53) and (7.54). $\qquad \square$

Case 2: $\left[Z, X \in H\right]$. In this case (7.39) reads

$$\nabla^{cr} dv(Z, X, \zeta) - \nabla^{cr} dv(X, Z, \zeta) = -R^{cr}(Z, X, \zeta, \nabla^{cr} v)$$
$$- \nabla^{cr} dv(T(Z, X), \zeta) = -2\Omega(Z, X)\nabla^{cr} dv(\zeta, \zeta), \quad (7.55)$$

where we used (7.2).

First we take a covariant derivative of (7.35) along $Z \in H$ and then substitute (7.34) and (7.35) in the obtained equality. Next, we anticommute the covariant derivatives and substitute the result in (7.55) together with the already established (7.41), (7.36), (7.20) and (7.38) to see, after some standard calculations, that the integrability condition in this case is

$$(\nabla_Z^{cr} \mathbb{D})(X, \zeta) - (\nabla_X^{cr} \mathbb{D})(Z, \zeta)$$
$$= -C(Z, JC(X)) + C(X, JC(Z)) - 2\mathbb{D}(\zeta, \zeta)\Omega(Z, X). \quad (7.56)$$

Lemma 7.3.7. *If the dimension of M is bigger than three and $S = 0$, then (7.56) holds.*

Proof. We differentiate the already proved (7.41) and then take the corresponding traces, using the symmetry of L, to see

$$((\nabla^{cr})^2_{e_a, Je_a} C)(Y, Z) - ((\nabla^{cr})^2_{e_a, Y} C)(Je_a, Z)$$
$$= (\nabla_Z^{cr} \mathbb{D})(Y, \zeta) + \Omega(Y, Z)(\nabla_{e_a}^{cr} \mathbb{D})(Je_a, \zeta) + 2(\nabla_Y^{cr} \mathbb{D})(Z, \zeta), \quad (7.57)$$

$$- ((\nabla^{cr})^2_{e_a, Y} C)(Je_a, Z) + ((\nabla^{cr})^2_{e_a, Z} C)(Je_a, Y)$$
$$= -(\nabla_Z^{cr} \mathbb{D})(Y, \zeta) + 2\Omega(Y, Z)(\nabla_{e_a}^{cr} \mathbb{D})(Je_a, \zeta) + (\nabla_Y^{cr} \mathbb{D})(Z, \zeta), \quad (7.58)$$

$$((\nabla^{cr})^2_{Y, e_a} C)(Je_a, Z) = -(2n + 1)(\nabla_Y^{cr} \mathbb{D})(Z, \zeta). \quad (7.59)$$

A combination of (7.57), (7.59) and (7.58) yields

$$\left[((\nabla^{cr})^2_{Y, e_a} C) - ((\nabla^{cr})^2_{e_a, Y} C)\right](Je_a, Z)$$
$$- \left[((\nabla^{cr})^2_{Z, e_a} C) - ((\nabla^{cr})^2_{e_a, Z} C)\right](Je_a, Y)$$
$$= 2n(\nabla_Z^{cr} \mathbb{D})(Y, \zeta) + \Omega(Y, Z)(\nabla_{e_a}^{cr} \mathbb{D})(Je_a, \zeta) - 2n(\nabla_Y^{cr} \mathbb{D})(Z, \zeta). \quad (7.60)$$

The Ricci identities, (7.2), (7.9), (7.47), (7.46) and (7.48) give

$$\left[((\nabla^{cr})^2_{Y, e_a} C) - ((\nabla^{cr})^2_{e_a, Y} C)\right](Je_a, Z) = 2(\nabla_\zeta^{cr} C)(Y, Z)$$
$$- (trC)[C(Y, JZ) + C(JY, Z)] + (2n + 1)C(Y, JC(Z)) + C(Z, JC(Y))$$
$$- 3C(JY, C(Z)) - 3C(JZ, C(Y)) - \Omega(Y, Z)C(e_a, JC(Je_a)). \quad (7.61)$$

$$((\nabla^{cr})^2_{e_a, Je_a} C)(Y, Z) = (n + 2)[C(JY, C(Z)) - C(Y, JC(Z))]$$
$$- (n + 2)C(Z, JC(Y)) + (n + 2)C(JZ, C(Y))$$
$$- 2n(\nabla_\zeta^{cr} C)(Y, Z) + (trC)(C(JY, Z) + C(Y, JZ)). \quad (7.62)$$

The identity (7.56) follows from (7.60) and (7.61). $\quad \square$

Case 3: $\left[X, Y \in H\right]$. In this case (7.39) reads

$$\nabla^{cr} dv(\zeta, X, Y) - \nabla^{cr} dv(X, \zeta, Y)$$
$$= -R^{cr}(\zeta, X, Y, \nabla^{cr} v) - \nabla^{cr} dv(T(\zeta, X), Y). \quad (7.63)$$

Take the covariant derivative of (7.34) along ζ and a covariant derivative of (7.35) along a horizontal direction, apply (7.35), (7.34), (7.36), use (7.7) and a suitable traces of (7.34) and (7.48) to get from (7.63) with the help of (7.48), (7.20) and the already proved (7.41) that the integrability condition (7.63) becomes

$$(\nabla^{cr}_X \mathbb{D})(Y, \zeta) - (\nabla^{cr}_\zeta C)(X, Y) =$$
$$C(Y, JC(X)) + A(X, C(Y)) + A(Y, C(X)) - \mathbb{D}(\zeta, \zeta)\Omega(X, Y). \quad (7.64)$$

Clearly Case 3 implies Case 2 since (7.56) is the skew-symmetric part of (7.64).

Lemma 7.3.8. *If the dimension of M is bigger than three and S = 0, then (7.64) holds.*

Proof. Combine (7.57), (7.59), (7.58) and the already proved (7.56) to obtain

$$((\nabla^{cr})^2_{e_a, Je_a} C)(Y, Z) + \left[((\nabla^{cr})^2_{Y, e_a} C) - ((\nabla^{cr})^2_{e_a, Y} C)\right](Je_a, Z)$$
$$= -2(n-1)(\nabla^{cr}_Y \mathbb{D})(Z, \zeta) + 2\Omega(Y, Z)\mathbb{D}(\zeta, \zeta) + C(Y, JC(Z))$$
$$- C(Z, JC(Y)) + \Omega(Y, Z)(\nabla^{cr}_{e_a} \mathbb{D})(Je_a, \zeta). \quad (7.65)$$

Now, (7.61), (7.62) and (7.65) imply (7.64). $\qquad\square$

Case 4: $\left[X \in H\right]$. In this case (7.39) has the form

$$\nabla^{cr} dv(X, \zeta, \zeta) - \nabla^{cr} dv(\zeta, X, \zeta) =$$
$$- R^{cr}(X, \zeta, \zeta, \nabla^{cr} v) + \nabla^{cr} dv(T(\zeta, X), \zeta) = A(X, \epsilon_i)\nabla^{cr} dv(\epsilon_i, \zeta). \quad (7.66)$$

We take the covariant derivative of (7.35) along ζ and a covariant derivative of (7.36) along a horizontal direction, then use (7.34) and the already proved (7.64). An application of (7.35) shows that (7.66) is equivalent to

$$(\nabla^{cr}_\zeta \mathbb{D})(X, \zeta) - (\nabla^{cr}_X \mathbb{D})(\zeta, \zeta) - 2\mathbb{D}(\epsilon_i, \zeta)C(X, J\epsilon_i)$$
$$+ A(X, \epsilon_i)\mathbb{D}(\epsilon_i, \zeta) = 0. \quad (7.67)$$

Lemma 7.3.9. *If the dimension of M is bigger than three and S = 0, then (7.67) holds.*

Proof. First we differentiate the already proven (7.56), (7.64), and the first equality in (7.37). Then we take the corresponding traces use the symmetry of C, A and (7.37) to obtain

$$((\nabla^{cr})^2_{e_a, Je_a}\mathbb{D})(Y,\zeta) = (n+2)[C(JY, e_b) - C(Y, Je_b)]\mathbb{D}(e_b, \zeta)$$
$$- (trC)\mathbb{D}(JY, \zeta) - 2n(\nabla_\zeta\mathbb{D})(Y,\zeta); \quad (7.68)$$

$$((\nabla^{cr})^2_{e_b, Je_b}\mathbb{D})(Y,\zeta) - ((\nabla^{cr})^2_{e_b,\zeta}C)(Je_b, Y)$$
$$= -(2n+1)\mathbb{D}(e_a, \zeta)[C(Y, Je_a) + A(Y, e_u)] + (\nabla^{cr}_Y\mathbb{D})(\zeta,\zeta)$$
$$+ [(\nabla^{cr}_{e_b}C)(Y, Je_a) + (\nabla^{cr}_{e_b}A)(Y, e_a)]C(Je_b, e_a)$$
$$+ (\nabla^{cr}_{e_b}A)(Je_b, e_a)C(Y, e_a) + A(Je_b, e_a)(\nabla^{cr}_{e_b}C)(Y, e_a) \quad (7.69)$$

$$((\nabla^{cr})^2_{\zeta, e_b}C)(Je_b, Y) = -(2n+1)(\nabla^{cr}_\zeta\mathbb{D})(Y,\zeta). \quad (7.70)$$

The Ricci identities, equations (7.7), (7.3), (7.4) and the symmetry of C imply

$$((\nabla^{cr})^2_{\zeta, e_b}C)(Je_b, Y) - ((\nabla^{cr})^2_{e_b,\zeta}C)(Je_b, Y)$$
$$= -(\nabla^{cr}_{Je_b}A)(e_b, e_a)C(Y, e_a) + A(e_b, Je_a)(\nabla^{cr}_{e_a}C)(e_b, Y)$$
$$+ [(\nabla^{cr}_{e_a}A)(e_b, Y) - (\nabla^{cr}_Y A)(e_b, e_a)]C(e_a, Je_b). \quad (7.71)$$

A small calculation taking into account (7.68), (7.69), (7.70), (7.71) and using (7.48) yields

$$(\nabla^{cr}_\zeta\mathbb{D})(Y,\zeta) - (\nabla^{cr}_Y\mathbb{D})(\zeta,\zeta) - 3[C(Y, Je_a) + A(Y, e_a)]\mathbb{D}(e_a, \zeta)$$
$$= -(trC)\mathbb{D}(JY, \zeta) + (\nabla^{cr}_{e_b}C)(Y, Je_a)C(Je_b, e_a)$$
$$+ [(\nabla^{cr}_{e_b}A)(Y, e_a) - (\nabla^{cr}_{e_a}A)(Y, e_b) + (\nabla^{cr}_Y A)(e_b, e_a)]C(Je_b, e_a). \quad (7.72)$$

Now, an application of the already proven (7.41) together with (7.48) to (7.72) yields the proof of (7.67). $\quad\square$

The proof of Theorem 7.3.5 i) is complete.

7.3.1 *The three dimensional case*

If the dimension is equal to 3 then it is easy to check that $S = 0$ and the integrability conditions (7.41) and (7.56) are trivially satisfied. Thus, the existence of a smooth solution depends only on the validity of (7.64) since the proof of Lemma 7.3.9 shows that (7.67) follows from (7.64) also in dimension three. The next Lemma 7.3.10 implies Theorem 7.3.5 ii).

Lemma 7.3.10. *If $n = 1$ and $F^{car} = 0$ then (7.64) holds.*

Proof. Suppose $n = 1$. Then identity (7.8) with $n = 1$ and (7.9) imply $r^{cr}(X, Y) = -\rho^{cr}(X, JY) = \frac{s^{cr}}{2}g(X, Y)$ which inserted into (7.20) yields

$$C(X, Y) = \frac{s^{cr}}{16}g(X, Y) + A(X, JY). \tag{7.73}$$

Apply (7.73) to (7.62) to get

$$2(\nabla_\zeta^{cr} C)(X, Y) = -((\nabla^{cr})^2_{e_a J e_a} C)(X, Y) - s^{cr}.A(X, Y). \tag{7.74}$$

The skew symmetric part of (7.64) is satisfied because $n = 1$. Now, (7.59), (7.74) and (7.73) give that the symmetric part of (7.64) is equivalent to $F^{car}(X, Y) = 0$. $\qquad \square$

The proof of Theorem 7.3.5 is complete. $\qquad \square$

Let us remark that the vanishing of F^{car} is equivalent to the vanishing of the Cartan curvature, cf. [Theorem 12.3, [161]]. As a consequence, we obtain [99]

Corollary 7.3.11. *A 3-dimensional Sasakian manifold* (M, θ, g, ζ) *is locally pseudoconformally equivalent to the three dimensional Heisenberg group if and only if its Riemannian scalar curvature* $Scal^g$ *satisfies*

$$(\nabla^g d(Scal^g))(X, JY) + (\nabla^g d(Scal^g))(Y, JX) = 0, \tag{7.75}$$

where ∇^g *is the Levi-Civita connection of* g.

Proof. It is well known that a pseudohermitian structure is Sasakian, i.e. its Riemannian cone is Kähler, exactly when the Webster torsion vanishes, $A = 0$. In particular, the Bianchi identities imply $\zeta(s^{cr}) = 0$. Then the terms in the second and third lines of (7.23) vanish in view of (7.19). On the other hand, for a Sasaki manifold, we have $\nabla_X^{cr} Y = \nabla_X^g Y + g(JX, Y)\zeta$ and the Riemannian scalar curvature and the scalar curvature of the Webster connection differ by an additive constant depending on the dimension, $2s^{cr} = Scal^g + 2n$ (see e.g. [60]). Now, (7.75) becomes equivalent to (7.23). Hence, (M, θ) is locally pseudoconformally flat. $\qquad \square$

Bibliography

[1] Adams, R. A. Sobolev spaces. Pure and Applied Mathematics, Vol. 65. Academic Press, New York-London, 1975. 12

[2] Alexandrov, A.D., *A characteristic property of the spheres*, Ann. Mat. Pura Appl. **58** (1962), 303–354. 64

[3] Alekseevsky, D. & Kamishima, Y., *Quaternionic and para-quaternionic CR structure on (4n+3)-dimensional manifolds*, Cent. Eur. J. Math. **2** (2004), no. 5, 732–753 (electronic). 131, 134

[4] Alekseevsky, D. & Kamishima, Y., *Pseudo-conformal quaternionic CR structure on (4n+3)-dimensional manifold*, Ann. Mat. Pura Appl. (4) **187** (2008), no. 3, 487–529., math.GT/0502531. 99, 154

[5] Alesker, S., *Non-commutative linear algebra and plurisubharmonic functions of quaternionic variables*, Bull. Sci. Math., **127** (2003), 1–35. v

[6] Alesker, S., *Quaternionic Monge-Ampére equations*, J. geom. anal., **13** (2003), 183–216. v

[7] Alesker, S., *Quaternionic plurisubharmonic functions and their applications to convexity*, math.DG/0606756. v

[8] Alesker, S. & Verbitsky, M., *Plurisubharmonic functions on hypercomplex manifolds and HKT-geometry*, J. Geom. Anal., **16** (2006), 375–399. v

[9] Alt, J., *Weyl connections and the local sphere theorem for quaternionic contact structures*, to appear in Annals of Global Analysis and Geometry. 155

[10] de Andres, L.C., Fernandez, M., Ivanov, St., Santisteban, J. A., Ugarte, L., & Vassilev, D., *Quaternionic Kaehler and Spin(7) metrics arising from quaternionic contact Einstein structures*, arXiv:1009.2745. 129, 137

[11] de Andres, L.C., Fernandez, M., Ivanov, St., Santisteban, J. A., Ugarte, L., & Vassilev, D., *Explicit Quaternionic Contact Structures and Metrics with Special Holonomy*, arXiv:0903.1398. 134, 135, 137

[12] Anselmi, D. & Fré, P., *Topological twist in four dimensions, R-duality and hyperinstantons*, Nuclear Phys. B 404 (1993), no. 1-2, 288–320. v

[13] Astengo, F., Cowling, M. & Di Blasio, B., *The Cayley transform and uniformly bounded representations*, J. Funct. Anal., **213** (2004), 241–269. 50

[14] Aubin, Th., *Problèms isopérimétriques et espaces de Sobolev*, J. Diff. Geometry, **11** (1976), 573–598. vi, vii, ix, 3, 88

207

[15] Aubin, Th., *Équations différentielles non linéaires et problème de Yamabe concernant la courbure scalaire*, J. Math. Pures Appl. (9) **55** (1976), no. 3, 269–296. vii

[16] Aubin, Th., *Espaces de Sobolev sur les variétés Riemanniennes*, Bull. Sc. Math., **100** (1976), 149–173. vi, vii, ix, 88

[17] Aubin, Th., & Cotsiolis, A., *Probléme de la courbure scalaire prescrite sur les variétés riemanniennes complètes*, J. Math. Pures Appl. (9) 81 (2002), no. 10, 999–1009. vii

[18] Badiale, M. & Tarantello, G., *A Sobolev-Hardy inequality with applications to a nonlinear elliptic equation arising in astrophysics*, Arch. Ration. Mech. Anal. **163** (2002), no. 4, 259–293. 88, 91

[19] Bahri, A., *Proof of the Yamabe conjecture, without the positive mass theorem, for locally conformally flat manifolds*. Einstein metrics and Yang-Mills connections (Sanda, 1990), 1–26, Lecture Notes in Pure and Appl. Math., 145, Dekker, New York, 1993. vii, 64

[20] Banner, A. D., *Some properties of boundaries of symmetric spaces of rank one*. Geom. Dedicata 88 (2001), no. 1-3, 113-133. 50

[21] Sub-Riemannian geometry. Edited by André Bellaïche and Jean-Jacques Risler. Progress in Mathematics, 144. Birkhuser Verlag, Basel, 1996. 5, 46

[22] Berndt, J., Tricerri, F., & Vanhecke, L., *Generalized Heisenberg groups and Damek-Ricci harmonic spaces*. Lecture Notes in Mathematics, 1598. Springer-Verlag, Berlin, 1995. 47

[23] Besse, A., Einstein manifolds, Ergebnisse der Mathematik, vol. 3, Springer-Verlag, Berlin, 1987. 137

[24] Biquard, O.,*Métriques d'Einstein asymptotiquement symétriques*, Astérisque **265** (2000). viii, x, xi, 97, 98, 99, 100, 102, 104, 105, 112, 117, 144

[25] Biquard, O., *Quaternionic contact structures*, Quaternionic structures in mathematics and phisics (Rome, 1999), 23–30 (electronic), Univ. Studi Roma "La Sapienza", Roma, 1999. 97, 98

[26] Birindelli, I., & Cutri, A., *A semilinear problem for the Heisenberg Laplacian*, Rend. Mat. Univ. Padova, **94** (1995), 137–153. 65

[27] Bonfiglioli, A., Lanconelli, E., & Uguzzoni, F. Stratified Lie groups and potential theory for their sub-Laplacians. Springer Monographs in Mathematics. Springer, Berlin, 2007. 5

[28] Boyer, Ch. & Galicki, K., *3-Sasakian manifolds*, Surveys in differential geometry: essays on Einstein manifolds, 123–184, Surv. Differ. Geom., **VI**, Int. Press, Boston, MA, 1999. 134

[29] Boyer, Ch., Galicki, K. & Mann, B., *The geometry and topology of 3-Sasakian manifolds* J. Reine Angew. Math., **455** (1994), 183–220. 131

[30] Branson, T. P., Fontana, L., Morpurgo, C., *Moser-Trudinger and Beckner-Onofris inequalities on the CR sphere*, arXiv:0712.3905v3. ix, 181, 183

[31] Brézis, H., & Kato, T.,*Remarks on the Schrödinger operator with singular complex potentials*, J. Math. Pures et Appl, 58 (1979), 137–151. 32

[32] Brézis,H., & Lieb, E., *A relation between pointwise convergence of functions and convergence of functionals*. Proc. Amer. Math. Soc. 88 (1983), no. 3, 486-490. 15

[33] Brézis, H., & Nirenberg, L., *Positive solutions of nonlinear elliptic equations involving critical Sobolev exponents*, Comm. Pure Appl. Math., 36 (1983), 437–477. 32

[34] Caffarelli, L., Kohn R., & Nirenberg L., *First Order Interpolation Inequality with Weights*, Compositio Math. 53 (1984) 259–275. xi, 89

[35] Capogna, L., & Cowling, M., *Conformality and Q-harmonicity in Carnot groups*. Duke Math. J. 135 (2006), no. 3, 455–479. 50

[36] Capogna, L., Danielli, D., Pauls, S. D., & Tyson, J. T., An introduction to the Heisenberg group and the sub-Riemannian isoperimetric problem. *Progress in Mathematics*, 259. Birkhuser Verlag, Basel, 2007. xi, 46

[37] Capogna, L., Danielli, D., & N. Garofalo, N., *Capacitary estimates and the local behavior of solutions of nonlinear subelliptic equations*, Amer. J. of Math., 118 (1997), 1153-1196. 55

[38] Capogna, L., Danielli, D., & Garofalo, N., *An imbedding theorem and the Harnack inequality for nonlinear subelliptic equations*, Comm. Part. Diff. Eq., 18 (1993), 1765–1794. 32, 38, 39

[39] Capria, M. & Salamon, S., *Yang-Mills fields on quaternionic spaces* Nonlinearity 1 (1988), no. 4, 517–530. 101

[40] Cartan, E., *Sur la geometrie pseudo-conforme des hypersurfaces de l'espace de deux variables complexes*,
I. Ann. di Mat., 11 (1932), 17–90,
II. Ann. Sci. Norm. Sup. Pisa, 1 (1932), 333–354. 191, 193, 196, 198

[41] Catrina, F., & Wang, Z-Q., *On the Caffarelli-Kohn-Nirenberg inequalities: sharp constants, existence (and nonexistence), and symmetry of extremal functions*. Comm. Pure Appl. Math. 54 (2001), no. 2, 229–258. 92

[42] Chen, W.,& Li, C., *Classification of solutions of some nonlinear elliptic equations*, Duke Math. Journ., 63 (1991), 615–622. 64, 72

[43] Chen, J. & Li, J., *Quaternionic maps between hyperkähler manifolds*, J. Diff. Geom. 55 (2000), no. 2, 355–384. v

[44] Chen, J. & Li, J., *Quaternionic maps and minimal surfaces*, Ann. Sc. Norm. Super. Pisa Cl. Sci. (5) 4 (2005), no. 3, 375–388. v

[45] Chern, S.S. & Moser, J., *Real hypersurfaces in complex manifolds*, Acta Math. 133 (1974), 219–271. 139, 191, 193, 195, 196, 197, 198

[46] Chow, W. L., *Über System von linearen partiellen Differentialgleichungen erster Ordnug*, Math. Ann., 117 (1939), 98-105. 8

[47] Ciatti, P., *A new proof of the J^2-condition for real rank one simple Lie algebras and their classification*. Proc. Amer. Math. Soc. 133 (2005), no. 6, 1611-1616. 47

[48] Corwin, L.J. & Greenleaf, F. P., Representations of nilpotent Lie groups and their applications. Part I. Basic theory and examples. Cambridge Studies in Advanced Mathematics, 18. Cambridge University Press, Cambridge, 1990. 5

[49] Cowling, M., Dooley, A. H., Korányi, A. & Ricci, F., *H-type groups and Iwasawa decompositions*, Adv. Math., 87 (1991), 1–41. viii, 41, 47, 48, 50, 52, 64

[50] Cowling, M., Dooley, A. H., Korányi, A. & Ricci, F., *An approach to symmetric spaces of rank one via groups of Heisenberg type.*, J. of Geom. Anal., **8** (1998), no. 2, 199–237. 41, 44, 47, 48, 50

[51] Cowling, M. & Korányi, A., *Harmonic analysis on Heisenberg type groups from a geometric viewpoint*, in "Lie Group Representation III", pp.60-100, Lec. Notes in Math., **1077** (1984), Springer-Verlag. 50, 52, 64

[52] Cowling, M., De Mari, F., Korányi, A., & H.M. Reimann, H.M, *Contact and conformal maps on Iwasawa N groups.* Atti Accad. Naz. Lincei Cl. Sci. Fis. Mat. Natur. Rend. Lincei (9) Mat. Appl. 13 (2002), 219–232. 50

[53] Cowling, M., De Mari, F., Korányi, A., & H.M. Reimann, H.M, *Contact and conformal maps in parabolic geometry. I*, Geom. Dedicata 111 (2005), 65–86. 50

[54] Cygan, J., *Subadditivity of homogeneous norms on certain nilpotent Lie groups*, Proc. Amer. Math. Soc., **83** (1981), 69-70. 8

[55] Damek, E., *Harmonic functions on semidirect extensions of type H nilpotent groups*, Trans. Amer. Math. Soc., 1, **290** (1985), 375–384. 68

[56] Damek, E., *The geometry of a semidirect extension of a Heisenberg type nilpotent group.* Colloq. Math. 53 (1987), no. 2, 255–268. 46

[57] Damek, E., & Ricci, F., *A class of nonsymmetric harmonic Riemannian spaces.* Bull. Amer. Math. Soc. (N.S.) 27 (1992), no. 1, 139–142. 47

[58] Damek, E., & Ricci, F., *Harmonic analysis on solvable extensions of H-type groups.* J. Geom. Anal. 2 (1992), no. 3, 213–248. 47, 48, 49

[59] De Mari, F., & Ottazzi, A., *Rigidity of Carnot groups relative to multicontact structures.* Proc. Amer. Math. Soc. 138 (2010), no. 5, 1889–1895. 50

[60] Dragomir, S. & Tomassini, G. Differential geometry and analisys on CR manifolds, Progress in Math. vol. 246, Birkhuser Boston, Inc., Boston, MA, 2006. 193, 205

[61] Duchemin, D., *Quaternionic contact structures in dimension 7*, Ann. Inst. Fourier (Grenoble) **56** (2006), no. 4, 851–885. 102, 103

[62] Eisenhart, L.P., Riemannian geometry, Princeton University Press, 1966. 191, 193

[63] Faraut, J., & Korányi, A., Analysis on symmetric cones. Oxford Mathematical Monographs. Oxford Science Publications. The Clarendon Press, Oxford University Press, New York, 1994. 49

[64] Fefferman, Ch. & Graham, C. R., *Conformal invariants. The mathematical heritage of Élie Cartan (Lyon, 1984).* Astérisque 1985, Numero Hors Serie, 95–116. 97

[65] Folland, G. B., *Subelliptic estimates and function spaces on nilpotent Lie groups*, Ark. Math., **13** (1975), 161–207. vii, 3, 5, 9, 36, 37

[66] Folland, G. B. *Applications of analysis on nilpotent groups to partial differential equations.* Bull. Amer. Math. Soc. 83 (1977), no. 5, 912–930. 38

[67] Folland, G. B.; Stein, Elias M., *Estimates for the $\bar{\partial}_b$ Complex and Analysis on the Heisenberg Group*, Comm. Pure Appl. Math., **27** (1974), 429–522. vii, 36, 37

[68] Folland, G. B., & Stein, Elias M. Hardy spaces on homogeneous groups. Mathematical Notes, 28. Princeton University Press, Princeton, N.J.; University of Tokyo Press, Tokyo, 1982. 5, 7

[69] Franchi, B., Gutierrez, C., & Wheeden, R. L., *Weighted Sobolev-Poincaré inequalities for Grushin type operators*, Comm. PDE, **19** (1994), 523–604. 10

[70] Frank, R. L., & Lieb, E.H., *Sharp constants in several inequalities on the Heisenberg group*, arXiv:1009.1410. viii, ix, 50, 88, 181, 183

[71] Fueter, R., *Über die analische Darstellung der regulären Funktionen einer Quaternionen variablen*, Comm. Math. Helv., **8** (1936), 371–378. 155

[72] Gamara, N., *The CR Yamabe conjecture the case $n = 1$*, J. Eur. Math. Soc. (JEMS) **3** (2001), no. 2, 105–137. vii

[73] Gamara, N. & Yacoub, R., *CR Yamabe conjecture – the conformally flat case*, Pacific J. Math. **201** (2001), no. 1, 121–175. vii

[74] Garofalo, N., & Nhieu, D. M., *Isoperimetric and Sobolev inequalities for Carnot-Carathéodory spaces and the existence of minimal surfaces*, Comm. Pure Appl. Math., **49** (1996), 1081–1144. 10, 33

[75] Garofalo, N. & Vassilev, D., *Symmetry properties of positive entire solutions of Yamabe type equations on groups of Heisenberg type*, Duke Math J, **106** (2001), no. 3, 411–449. xi, 88, 89, 158

[76] Garofalo, N. & Vassilev, D., *Regularity near the characteristic set in the non-linear Dirichlet problem and conformal geometry of sub-Laplacians on Carnot groups*, Math Ann., **318** (2000), no. 3, 453–516. ix, xi, 56, 64

[77] Ghoussoub, N. & Yuan, C., *Multiple solutions for quasi-linear PDEs involving the critical Sobolev and Hardy exponents*, Trans. Amer. Math. Soc. **352** (2000), no. 12, 5703–5743. xi, 89

[78] Gidas, B., Ni, W. M. & Nirenberg, L., *Symmetry and related properties via the maximum principle*, Comm. Math. Phys., **68** (1979), 209–243. ix, 64

[79] Gidas, B., Ni, W. M. & Nirenberg, L., *Symmetry of positive solutions of nonlinear elliptic equations*, Math. Anal. Appl., Part A, Adv. in Math. Suppl. Studies, **7 A** (1981), 369–402. 64

[80] Gilbarg, D., & Trudinger, N. S., Elliptic partial differential equations of second order, Second ed., Revised third printing, Vol. 244, Springer-Verlag, Grundlehren der mathematischen Wissenschaften, 1998. 65, 67

[81] Glaser, V., Martin, A., Grosse, H., & Thirring, W., *A family of optimal conditions for absence of bound states in a potential*, In: Studies in Mathematical Physics. Lieb, E.H., Simon, B., Wightman, A.S. (eds.), pp. 169–194. Princeton University Press 1976 89

[82] Goodman, Roe W., Nilpotent Lie groups: structure and applications to analysis. Lecture Notes in Mathematics, Vol. 562. Springer-Verlag, Berlin-New York, 1976. 5

[83] Graham, C. R. & Lee, J. M., *Einstein metrics with prescribed conformal infinity on the ball*, Advances in Math. **87** (1991), 186–225. 97

[84] Gray, A., *Einstein manifolds which are not Einstein,* Geom. Dedicata **7** (1978), 259-280. 134

[85] Gromov, M., *Carnot-Carathodory spaces seen from within*. Sub-Riemannian geometry, 79–323, Progr. Math., 144, Birkhuser, Basel, 1996. xi

[86] Harvey, F. R., & Lawson Jr, H. B., *Plurisubharmonic Functions in Calibrated Geometries*, preprint, math.CV/0601484. v

[87] Hebisch, W., & Sikora, A., *A smooth subadditive homogeneous norm on a homogeneous group*. Studia Math. 96 (1990), no. 3, 231–236. 8

[88] Helgason, S., Differential geometry, Lie groups, and symmetric spaces. Pure and Applied Mathematics, 80. Academic Press, Inc. [Harcourt Brace Jovanovich, Publishers], New York-London, 1978. 44

[89] Hersch, J., *Quatre propriétés isopérimétriques de membranes sphériques homogénes.* C. R. Acad. Sci. Paris Sr. A-B **270** (1970), A1645A1648. ix, 159, 182

[90] Hitchin, N., *The self-duality equations on a Riemann surface,* Proc. London Math. Soc. **55** (1987) 59–126. 132

[91] Holopainen, I., *Positive solutions of quasilinear elliptic equations on Riemannian manifolds,* Proc. London Math. Soc., (3) **65** (1992), no. 3, 651–672. 32, 67

[92] I. Holopainen & S. Rickman, *Quasiregular mappings of the Heisenberg group,* Math. Ann., **294** (1992), no. 4, 625-643 32

[93] Hörmander, L., *Hypoelliptic second-order differential equations,* Acta Math., **119** (1967), 147–171. 36

[94] Ivanov, S., Minchev, I., & Vassilev, D., *Quaternionic contact Einstein structures and the quaternionic contact Yamabe problem.* preprint, math.DG/0611658. v, xii, 98, 112, 113, 119, 124, 127, 130, 133, 135, 139, 140, 142, 144, 145, 155, 159, 163, 165, 174, 182

[95] Ivanov, S., Minchev, I., & Vassilev, D., *Extremals for the Sobolev inequality on the seven dimensional quaternionic Heisenberg group and the quaternionic contact Yamabe problem,* J. Eur. Math. Soc. (JEMS) 12 (2010), no. 4, 1041–1067. viii, ix, xii, 5, 159

[96] Ivanov, S., Minchev, I., & Vassilev, D., *The optimal constant in the L^2 Folland-Stein inequality on the quaternionic Heisenberg group,* arXiv:1009.2978 viii, ix, xii, 5, 88, 158

[97] Ivanov, S., D. Vassilev, *Conformal quaternionic contact curvature and the local sphere theorem,* J. Math. Pures Appl. **93** (2010), pp. 277-307. xii, 113, 119, 127, 137, 147, 150, 151, 152, 153, 154, 191

[98] Ivanov, S., D. Vassilev, *Quaternionic contact manifolds with a closed fundamental 4-form,* Bull. London Math. Soc. 2010; doi:10.1112/blms/bdq061; arXiv:0810.3888 . 117, 118, 130, 155

[99] Ivanov, S., D. Vassilev, Zamkovoy, S., *Conformal Paracontact curvature and the local flatness theorem,* Geom. Dedicata **144** (2010), 79-100. xii, 191, 205

[100] Jelonek, W., *Positive and negative 3-K-contact structures,* Proc. Am. Mat. Soc. **129** (2000), 247-256. 131, 134

[101] Jerison, D., & Lee, J., *A subelliptic, nonlinear eigenvalue problem and scalar curvature on CR manifolds,* Contemporary Math., **27** (1984), 57–63. vii

[102] Jerison, D., & Lee, J., *The Yamabe problem on CR manifolds,* J. Diff. Geom.,**25** (1987), 167–197. 64, 78, 84

[103] Jerison, D., & Lee, J., *Extremals for the Sobolev inequality on the Heisenberg group and the CR Yamabe problem,* J. Amer. Math. Soc., **1** (1988), no. 1, 1–13. ix, 5, 56, 88

[104] Jerison, D., & Lee, J., *Intrinsic CR normal coordinates and the CR Yamabe problem,* J. Diff. Geom., **29** (1989), no. 2, 303–343. vii

[105] Kaplan, A., *Fundamental solutions for a class of hypoelliptic PDE generated by composition of quadratic forms*, Trans. Amer. Math. Soc., **258** (1980), 147–153. 41, 42, 44, 45

[106] Kaplan, A., *Riemannian nilmanifolds attached to Clifford modules*. Geom. Dedicata 11 (1981), no. 2, 127-136. 42

[107] Kaplan, A., & Putz, R., *Harmonic forms and Riesz transforms for rank one symmetric spaces*. Bull. Amer. Math. Soc. 81 (1975), 128-132. 56

[108] Kaplan, A., & Putz, R., *Boundary behavior of harmonic forms on a rank one symmetric space*. Trans. Amer. Math. Soc. 231 (1977), no. 2, 369-384. 43, 47, 63

[109] Kaplan, A., & Ricci, F., *Harmonic analysis on groups of Heisenberg type*. Harmonic analysis (Cortona, 1982), 416435, Lecture Notes in Math., 992, Springer, Berlin, 1983. 47

[110] Knapp, A. W., & Stein, E. M., *Intertwining operators for semisimple groups*. Ann. of Math. (2) 93 (1971), 489-578. 7, 8

[111] Kobayashi, S., *Principal fibre bundles with 1-dimensional toroidal group*, Tohoku Math. J. (56) **8** (1956), 29–45. 137

[112] Kobayashi, Sh., & Nomizu, K., *Foundations of differential geometry. Vol. II.* Interscience Tracts in Pure and Applied Mathematics, No. 15 Vol. II Interscience Publishers John Wiley & Sons, Inc., New York-London-Sydney 1969 xv+470 pp. 105

[113] Konishi, M., *On manifolds with Sasakian 3-structure over quaternion Kähler manifolds*, Kodai Math. Sem. Rep. **26** (1975), 194-200. 131, 134

[114] Korányi, A., *Kelvin transform and harmonic polynomials on the Heisenberg group*, Adv.Math. **56** (1985), 28–38. 50, 51, 64

[115] Korányi, A., *Geometric properties of Heisenberg-type groups*. Adv. in Math. 56 (1985), no. 1, 28-38. 43, 46, 47

[116] Korányi, A., *Multicontact maps: Results and conjectures*. Lect. Notes Semin. Interdiscip. Mat., IV, S.I.M. Dep. Mat. Univ. Basilicata, Potenza, 2005, 57–63. 50

[117] Korányi, A., & Vági, S., *Singular integrals on homogeneous spaces and some problems of classical analysis*. Ann. Scuola Norm. Sup. Pisa (3) 25 (1971), 575-648 (1972). 7

[118] Krantz, S., *Lipschitz spaces on stratified groups*, Trans. Amer. Math. Soc., **269** (1982), no. 1, 39–66. 37

[119] Lanconelli, E., & Uguzzoni, F., *Asymptotic behavior and non-existence theorems for semilinear Dirichlet problems involving critical exponent on unbounded domains of the Heisenberg group*, Boll. Un. Mat. Ital., (8) **1-B** (1998), 139–168. 32, 41

[120] Lee, J., *Pseudo-einstein structures on CR manifolds*, Amer. J. Math., **110** (1988), 157–178. 193

[121] Lee, J., *The Fefferman metric and pseudohermitian invariants*, Trans. AMS, **296** (1986), 411–429. 192, 194, 197

[122] Lee, J. M. & Parker, T., *The Yamabe Problem*, Bull Am. Math. Soc. **17** (1987), no. 1, 37–91. vii

[123] Li, J. & Zhang, X., *Quaternionic maps between quaternionic Kähler manifolds*, Math. Z. **250** (2005), no. 3, 523–537. v

[124] Lin, C-S, & Wang, Z-Q., *Symmetry of extremal functions for the Caffarelli-Kohn-Nirenberg inequalities.* Proc. Amer. Math. Soc. **132** (2004), no. 6, 1685–1691. 92

[125] Lions, P.L., *The concentration compactness principle in the calculus of variations. The limit case. Part 1*, Rev. Mat. Iberoamericana **1.1** (1985), 145–201. 4, 19

[126] Lions, P.L., *The concentration compactness principle in the calculus of variations. The limit case. Part 2*, Rev. Mat. Iberoamericana **1.2** (1985), 45–121. 4, 28

[127] Mancini, G., Fabbri, I., & Sandeep, K., *Classification of solutions of a critical Hardy-Sobolev operator.* J. Differential Equations 224 (2006), no. 2, 258–276. 89, 91

[128] Maz'ja, V. G. Sobolev spaces. Translated from the Russian by T. O. Shaposhnikova. Springer Series in Soviet Mathematics. Springer-Verlag, Berlin, 1985. 12

[129] Mancini, G. & Sandeep, K., *Cylindrical symmetry of extremals of a Hardy–Sobolev inequality.*, Ann. Mat. Pura Appl. (4) **183** (2004), no. 2, 165–172 92

[130] Mitchell, J., *On Carnot-Carathéodory metrics.* J. Differential Geom. 21 (1985), no. 1, 3545. 9

[131] Moser, J., *On Harnack's theorem for elliptic differential equations,* Comm. Pure Appl. Math.14 (1961), 577–591. 32, 34

[132] Nagel, Q., E. M. Stein, E.M., & Wainger, S., *Balls and metrics defined by vector fields I: basic properties,* Acta Math. **155** (1985), 103–147. 5

[133] Obata, M., *The conjectures on conformal transformations of Riemannian manifolds,* J. Differential Geometry **6** (1971/72), 247–258. ix

[134] Okikiolu, G. O. , *Aspects of the theory of bounded integral operators in L^p-spaces,* Academic Press, London, 1971. 89

[135] O'Neill, B., Semi-Riemannian geometry, Academic Press, New York, 1883.

[136] Pansu, P., *Métriques de Carnot-Carathéodory et quasiisométries des espaces symétriques de rang un,* Ann. of Math. (2) 129 (1989), no. 1, 1–60. 97

[137] Pansu, P., *Quasiisométries des variétés á courbure négative.* Thése, Paris 1987. 50

[138] Pertici, D., *Funzioni regolari di piú variabili quaternioniche,* Ann. Mat. Pura Appl., Serie IV, **CLI** (1988), 39–65. v

[139] Pertici, D., *Trace theorems for regular functions of several quaternionic variables,* Forum Math. 3 (1991), 461–478. v

[140] Quillen, D., *Quaternionic algebra and sheaves on the Riemann sphere,* Quart. J. Math. Oxford, **49** (1998), 163–198. v

[141] Rashevsky, P. K., *Any two points of a totally nonholonomic space may be connected by an admissible line,* Uch. Zap. Ped. Inst. im. Liebknechta, Ser. Phys. Math., (Russian) **2** (1938), 83–94. 8

[142] Reimann, H.M., *Rigidity of H-type groups.* Math. Z. 237 (2001), no. 4, 697–725. 50

[143] Riehm, C. *The automorphism group of a composition of quadratic forms.* Trans. Amer. Math. Soc. 269 (1982), no. 2, 403–414. 42

[144] Rothschild, L. P., & Stein, E. M., *Hypoelliptic differential operators and nilpotent groups.* Acta Math. **137** (1976), 247–320. 36

[145] Reed, M., & Simon, B., Methods of modern mathematical physics. I. Functional analysis. Second edition. Academic Press, Inc., New York, 1980. 11, 13

[146] Rudin, W., Real and complex analysis. Third edition. McGraw-Hill Book Co., New York, 1987. 14

[147] Salamon, S., *Almost parallel structures* Global differential geometry: the mathematical legacy of Alfred Gray (Bilbao, 2000), 162–181, Contemp. Math., 288, Amer. Math. Soc., Providence, RI, 2001. 137

[148] Saloff-Coste, Laurent, Aspects of Sobolev-type inequalities. London Mathematical Society Lecture Note Series, 289. Cambridge University Press, Cambridge, 2002. xi

[149] Schoen, R., *Conformal deformation of a Riemannian metric to constant scalar curvature,* J. Differential Geom., **20** (1984), no. 2, 479–495. vii

[150] Secchi, S., Smets, D. & Willem, M., *Remarks on a Hardy-Sobolev inequality,* C. R. Math. Acad. Sci. Paris **336** (2003), no. 10, 811–815. 92

[151] Serrin, J., *A symmetry problem in potential theory,* Arch. Ration. Mech., **43** (1971), 304–318. 64

[152] Serrin, J., *Local behavior of solutions of quasi-linear equations,* Acta Math., **111** (1964), 247-302 32

[153] Stein, E. M., Harmonic analysis: real-variable methods, orthogonality, and oscillatory integrals. With the assistance of Timothy S. Murphy. Princeton Mathematical Series, 43. Monographs in Harmonic Analysis, III. Princeton University Press, Princeton, NJ, 1993. 48

[154] Stiefel, E., *On Cauchy-Riemann Equations in Higher Dimensions,* J. Res. Nat. Bureau of Standards, **48** (1952), 395–398. v

[155] Struwe, M., *Variational Methods, Aplications to Nonlinear Partial Differential Equations and Hamilton Systems,* Fourth edition. Springer-Verlag, Berlin, 2008. 19

[156] Sudbery, A., *Quaternionic Analysis,* Math. Proc. Camb. Phil. Soc., **85** (1979), 199-225. v

[157] Swann, A., *Quaternionic Kähler geometry and the fundamental 4-form,* Proceedings of the Workshop on Curvature Geometry (Lancaster, 1989), 165–173, ULDM Publ., Lancaster, 1989. 137

[158] Szegö, G., *Inequalities for certain eigenvalues of a membrane of given area* J. Rational Mech. Anal. 3, (1954). 343–356. ix, 159, 182

[159] Talenti, G., *Best constant in the Sobolev inequality,* Ann. Mat. Pura Appl., **110** (1976), 353–372. vi, ix, 3, 88

[160] Tanaka, N., *A differential geometric study on strongly pseudo-convex manifolds,* Lectures in Mathematics, Department of Mathematics, Kyoto University, No. 9. Kinokuniya Book-Store Co., Ltd., Tokyo, 1975. 9, 97, 192

[161] Tanaka, N., *On non-degenerate real hypersurfaces, graded Lie algebras and Cartan connections,* Japan J. Math. 2 (1976), 131-190. 205

[162] Tanno, S., *Remarks on a triple of K-contact structures,* Tohoku Math. J. **48** (1996), 519-531.

[163] Trudinger, N., *Remarks concerning the conformal deformation of Riemannian structures on compact manifolds*, Ann. Scuola Norm. Sup. Pisa, **22** (1968), 265–274. 131, 134

[164] Varopoulos, N. Th.; Saloff-Coste, L.; Coulhon, T. Analysis and geometry on groups. Cambridge Tracts in Mathematics, 100. Cambridge University Press, Cambridge, 1992. vi

[165] Vassilev, D., *Yamabe type equations on Carnot groups*, Ph. D. thesis Purdue University, 2000. xi

[166] Vassilev, D., *Regularity near the characteristic boundary for sub-laplacian operators*, Pacific J Math, **227** (2006), no. 2, 361–397. xi

[167] Vassilev, D., *L^p estimates, asymptotic behavior and extremals for some Hardy-Sobolev inequalities*, to appear in Trans. AMS. xi, 16
xi, 41, 89, 91

[168] Wang, W., *The Yamabe problem on quaternionic contact manifolds*, Ann. Mat. Pura Appl., **186** (2006), no. 2, 359–380. x, 140, 162

[169] Webster, S. M., *Real hypersurfaces in complex space*, Thesis, University of California, 1975. 64, 84, 97, 139, 191, 192, 193, 198

[170] Webster, S. M., *Pseudo-hermitian structures on a real hypersurface*, J.Diff. Geom., **13** (1979), 25–41. 192, 193

[171] Wolf, J.A., Spaces of constant curvature. Second edition. Department of Mathematics, University of California, Berkeley, Calif., 1972. 44

[172] Wu, H.-H., *The Bochner technique in differential geometry*, Math. Rep., **3** (1988), no. 2, i–xii and 289–538. 160

[173] Xu, C. J., *Subelliptic variational problems*, Bull. Soc. Math. France, **118** (1990), 147–169. 32

[174] Yamabe, H. , *On a deformation of Riemannian structures on compact manifolds*, Osaka Math J., (12) (1960), 21–37. v, vi

[175] Yosida, K., Functional analysis. Reprint of the sixth (1980) edition. Classics in Mathematics. Springer-Verlag, Berlin, 1995. 11, 14

Index